Adenosine and Adenine Nucleotides: Physiology and Pharmacology

Adenosine and Adenine Nucleotides: Physiology and Pharmacology

Edited by
David M. Paton
University of Auckland,
New Zealand

Taylor & Francis
London, Philadelphia and New York
1988

UK Taylor & Francis Ltd, 4 John St, London WC1N 2ET

USA Taylor & Francis Inc., 242 Cherry St, Philadelphia, PA 19106-1906

British Library Cataloguing in Publication Data

Adenosine and adenine nucleotides:
physiology and pharmacology.
1. Adenylic acid. 2. Adenosine
3. Pharmacology, Experimental
I. Paton, David M. (David Murray)
615′.1 QP951.A25

ISBN 0-85066-416-0

Printed in Great Britain by Taylor & Francis (Printers) Ltd. Basingstoke, Hants.

CONTENTS

CLASSIFICATION AND PROPERTIES OF ADENOSINE RECEPTORS

RECEPTORS FOR ADENINE NUCLEOTIDES

MECHANISMS OF ACTION OF ADENOSINE

ACTIONS OF ADENOSINE ON THE CARDIOVASCULAR SYSTEM

PREFACE

A Satellite Symposium on the "PHYSIOLOGY AND PHARMACOLOGY OF ADENOSINE AND ADENINE NUCLEOTIDES" was held at the School of Medicine, University of Auckland in Auckland, New Zealand, from 31 August - 2 September 1987, immediately following the 10th International Pharmacology Congress in Sydney. This Symposium was attended by investigators from Asia, Europe, North America and Australasia. The present volume contains the invited papers presented at this meeting, together with the abstracts of poster presentations.

It is now nearly 60 years since Drury and Szent-Gyorgyi published their classic paper in which they described the cardiac effects of adenosine and related compounds. In arranging the programme for this Satellite Symposium, particular attention was paid to ensuring that coverage was provided for areas of recent study and development. Another aim was to provide a forum for the presentation of different aspects of the physiology and pharmacology of adenosine and adenine nucleotides work from as many laboratories as possible.

I wish to thank the participants for the careful preparation of their presentations and for their participation in this Symposium. The Symposium could not have been held without the financial and other support provided by a wide variety of philanthropic and business organisations. These are listed in the acknowledgement section. I am pleased to acknowledge the detailed administrative and secretarial assistance provided for the organisation of the Symposium by Mrs Jill Simpson and the careful secretarial assistance provided by Mrs Dorothy Vance and Mrs Zea Eagle in the final preparation of this volume. I am also grateful to Dr John Cheney of Taylor & Francis Limited for his advice in relation to the publication of these proceedings.

David M Paton

Auckland
October, 1987

ACKNOWLEDGEMENT

This Satellite Symposium was made possible by grants, financial support and other assistance provided by the Foundations and Organisations listed below, and this support is gratefully acknowledged.

Auckland Medical Research Foundation

Butland Medical Foundation

Department of Pharmacology and Clinical Pharmacology, University of Auckland

Medical Research Council of New Zealand

Sir John Logan Campbell Medical Trust

Air New Zealand

Bank of New Zealand

Beecham Research Laboratories

Boehringer-Ingelheim (New Zealand) Limited

Ciba-Geigy Corporation

Glaxo (New Zealand) Limited

Merck Sharp and Dohme (New Zealand) Limited

Pharmaco (New Zealand) Limited

Research Biochemicals Inc.

Roche Products (New Zealand) Limited

E R Squibb and Sons (New Zealand) Limited

Upjohn Inter-American Corporation (New Zealand)

Warner-Lambert (Parke-David) New Zealand Limited

CONTRIBUTORS

R.A. Barraco
Department of Physiology
Wayne State University Medical School
Detroit, Michigan 48201
UNITED STATES OF AMERICA

J-M Boeynaems
Institute of Interdisciplinary Research
School of Medicine
Universite Libre de Bruxelles
1070 - Brussels
BELGIUM

J.A. Bristol
Parke-Davis Pharmaceutical Research Division
Warner-Lambert Company
Ann Arbor, Michigan 48105,
UNITED STATES OF AMERICA

R.F. Bruns
Department of Pharmacology
Parke-Davis Pharmaceutical Research Division
Warner-Lambert Company
Ann Arbor, Michigan 48105
UNITED STATES OF AMERICA

M.G. Collis
Bioscience II Department
ICI Pharmaceuticals Division
Alderley Park
Macclesfield, Cheshire SK10 4TG
UNITED KINGDOM

N.J. Cusack
Department of Pharmacology
King's College London
Strand
London WC2R 2LS
UNITED KINGDOM

T.V. Dunwiddie
Department of Pharmacology
University of Colorado Health Sciences Center
Denver, Colorado 80262
UNITED STATES OF AMERICA

R.W. Foster
Department of Physiological Sciences
University of Manchester
Manchester M13 9PT
UNITED KINGDOM

B.B. Fredholm
Department of Pharmacology
Karolinska Institutet
10401 Stockholm
SWEDEN

K.A. Jacobson
Laboratories of Chemistry and Bioorganic Chemistry
National Institutes of Health
Bethesda, Maryland 20892
UNITED STATES OF AMERICA

W.W. Lautt
Department of Pharmacology and Therapeutics
University of Manitoba
Winnipeg, Manitoba
CANADA R3E OW3

H. Osswald
Goedecke Research Institute
7800 Freiburg
WEST GERMANY

J.R. Parratt
Department of Physiology and Pharmacology
University of Strathclyde
Royal College
Glasgow G1 1XW
UNITED KINGDOM

D.M. Paton
Department of Pharmacology and Clinical Pharmacology
University of Auckland
Auckland
NEW ZEALAND

A. Pelleg
The Lankenau Medical Research Center
Philadelphia, Pennsylvania 19151
UNITED STATES OF AMERICA

C.G.A. Persson
Department of Clinical Pharmacology
University Hospital
S-221 85 Lund
SWEDEN

J.A. Ribeiro
Laboratory of Pharmacology
Gulbenkian Institute of Science
2781 Oeiras
PORTUGAL

D. Robertson
Departments of Medicine and Pharmacology
Vanderbilt University
Nashville, Tennessee 37232
UNITED STATES OF AMERICA

D. Satchell
Department of Zoology
University of Melbourne
Parkville
Victoria 3052
AUSTRALIA

W. Schmitz
Department of Pharmacology
University of Hamburg
D-2000 Hamburg 20
FEDERAL REPUBLIC OF GERMANY

P. Schubert
Department of Neuromorphology,
Max Planck Institute for Psychiatry
8033 Planegg-Martinsried,
FEDERAL REPUBLIC OF GERMANY

U. Schwabe,
Department of Pharmacology
University of Heidelberg
6900 Heidelberg
FEDERAL REPUBLIC OF GERMANY

G.L. Stiles
Departments of Medicine and Biochemistry
Duke University Medical Center
Durham, North Carolina 27710
UNITED STATES OF AMERICA

T.W. Stone
Department of Physiology
St. George's Medical School
University of London
London SW 17 ORE
UNITED KINGDOM

H. Van Belle
Department of Biochemistry
Janssen Pharmaceutica
B-2340 Beerse
BELGIUM

T.D. White
Department of Pharmacology
Dalhousie University
Halifax, Nova Scotia,
CANADA B3H 4H7

CLASSIFICATION AND PROPERTIES OF ADENOSINE RECEPTORS

USE OF RADIOLIGANDS IN THE STUDY OF ADENOSINE RECEPTORS

ULRICH SCHWABE

Pharmakologisches Institut der Universität Heidelberg
Im Neuenheimer Feld 366, 6900 Heidelberg
Federal Republic of Germany

INTRODUCTION

During the last two decades important advances have been made in our understanding of the actions of adenosine. These became possible when metabolically stable adenosine analogs were synthetized which facilitated the study of physiological adenosine effects. Furthermore, well known drugs of the group of methylxanthines were found to act as adenosine antagonists and provided additional pharmacological tools in adenosine research. An important step in elucidating the biological role of adenosine was the proposal of membrane-bound adenosine receptors derived from adenylate cyclase studies (Sattin and Rall 1970).

Two types of extracellular adenosine receptors were originally distinguished which are both linked to adenylate cyclase (van Calker et al. 1978, Londos et al. 1980). The A_1 adenosine receptor has a relatively high affinity for adenosine and mediates inhibition of adenylate cyclase activity (Table 1). The A_2 adenosine receptor has generally a lower affinity and mediates stimulation of the enzyme activity. Both receptor subtypes were also defined by their relative affinities for the agonists R-N^6-phenylisopropyladenosine (R-PIA) and 5′-N-ethylcarboxamidoadenosine (NECA). A_1 receptors have a higher affinity for R-PIA than for NECA while A_2 adenosine receptors show the reverse potency order (Londos et al. 1980). Recently, a further subdivision of A_2 receptors has been proposed into A_{2a} receptors with high affinity and A_{2b} receptors with low affinity for NECA, since the affinity of A_2 receptors for NECA was found to be markedly different in several tissues (Daly et al. 1983, Bruns et al. 1986). In addition to these external receptors a so-called purine site (P-site) for

Table 1. Classification of adenosine receptor subtypes

Subtype	Effect	Agonists
A_1	Inhibition of AC	R-PIA > NECA
A_1	Opening of potassium channels	R-PIA
A_{2a}	Stimulation of AC in striatum	NECA > R-PIA N^6-Methyladenosine > MPA[+]
A_{2b}	Stimulation of AC in fibroblasts	NECA > R-PIA MPA > N^6-Methyladenosine

[+]2-(4-Methoxyphenyl)adenosine

adenosine has been localized to the internal site of the cell membrane mediating the inhibitory effects of high adenosine concentrations on the catalytic site of the adenylate cyclase.

Furthermore, evidence has been obtained for cyclase-independent adenosine receptors in the heart. Atrial A_1 adenosine receptors appear to be coupled to potassium channels which are opened by adenosine agonists without corresponding changes in cAMP levels (Böhm et al. 1986).

Shortly after the adenosine receptor concept was proposed these receptors were directly probed in binding experiments with radioactively labeled adenosine agonists and antagonists. In the following years a growing number of radioligands has been developed for A_1 and A_2 receptors. The majority of the ligands were selective for A_1 receptors whereas the analysis of A_2 receptors turned out to be more difficult. The aim of this chapter is to review the recently acquired information on the characterization of both adenosine receptor subtypes by radioligand binding.

RADIOLIGANDS FOR A_1 ADENOSINE RECEPTORS

Radiolabeled agonists

A successful identification of A_1 adenosine receptors was achieved with several radiolabeled agonists of high affinity and specifity such as $[^3H]N^6$-cyclohexyladenosine ($[^3H]$CHA) (Bruns et al. 1980), $[^3H]$R-N^6-phenylisopropyladenosine ($[^3H]$PIA) (Schwabe and Trost 1980) and $[^3H]N^6$-cyclopentyladenosine ($[^3H]$CPA) (Williams et al. 1986). Less

Table 2. Agonist radioligands for A_1 adenosine receptors

Radioligand	Rat brain K_D (nM)	Nonspecific binding at K_D	Specific activity Ci/mmol
[^{3}H]ClA	1.3	5 - 10 %	12
[^{3}H]CHA	0.9	6 %	25
[^{3}H]PIA	1.4	5 %	50
[^{3}H]CPA	0.48	5 - 10 %	46
[^{125}I]HPIA	0.48	15 %	2,175

satisfactory results were obtained with [^{3}H]2-chloroadenosine because of the probable instability of the ligand (Williams and Risley 1980, Haubrich et al. 1981). All these adenosine analogs are stable against adenosine metabolizing enzymes and thus fulfill a general requirement for radioligands. In addition, iodinated radioligands have been developed such as [^{125}I]R-N^6-p-hydroxyphenylisopropyladenosine ([^{125}I]HPIA) (Schwabe et al. 1982), [^{125}I]N^6-p-aminobenzyladenosine ([^{125}I]ABA) (Linden et al. 1985) and [^{125}I]N^6-2-(4-aminophenyl)ethyladenosine ([^{125}I]APNEA) (Stiles et al. 1985). Iodinated ligands have much higher specific radioactivity and can therefore be used for the detection of A_1 receptors in tissues with low receptor density. Additional data of the agonist radioligands are listed in Table 2.

[^{3}H]CHA and [^{3}H]PIA have been most widely used for the characterization of A_1 receptors in several tissues of various species (Schwabe 1985).

The pharmacological profile of radioligand binding to A_1 adenosine receptors is determined by competition experiments with adenosine agonists and antagonists. The binding profile obtained with rat brain membranes is almost identical with all three N^6-substituted radioligands in different laboratories (Table 3).

CPA is always the most potent agonist with K_i-values of 0.6 - 0.8 nM. R-PIA and CHA are slightly less potent than CPA, whereas NECA as the prototypical A_2 receptor agonist is about 10-fold less potent than CPA. S-PIA is 30-fold less potent than R-PIA demonstrating a high degree of stereoselectivity of binding.

Among the adenosine antagonists 8-cyclopentyl-1,3-dipropylxanthine (DPCPX) is the most potent compound with a K_i

Table 3. Pharmacological profile of A_1 adenosine receptors. Radioligand binding to rat brain membranes.
Data from [a]Lohse et al. (1984 and 1987), [b]Bruns et al. (1986), [c]Williams et al. (1986).

Compound	$[^3H]PIA$[a]	$[^3H]CHA$[b]	$[^3H]CPA$[c]
		K_i (nM)	
Agonists			
CPA	0.8	0.6	0.6
R-PIA	1.3	1.2	1.3
CHA	2.6	1.3	1.4
NECA	8.2	6.3	5.1
2-Chloroadenosine	14.4	9.3	6.5
S-PIA	31.0	49.3	21.5
Antagonists			
DPCPX	0.3		
DPX	70	44.5	
8-Phenyltheophylline	115	86	82
IBMX	3,170	2,460	1,870
Theophylline	11,000	8,470	5,140

of 0.3 nM followed by two other 8-phenyl-substituted xanthine derivatives, 3-isobutyl-1-methylxanthine (IBMX) and the classical methylxanthine theophylline.

Radiolabeled antagonists

All radiolabeled antagonists for A_1 adenosine receptors originate from the class of xanthine derivatives. $[^3H]$1,3-diethyl-8-phenylxanthine ($[^3H]$DPX) was introduced as the first antagonist radioligand (Bruns et al. 1980). 8-Cyclopentyl-1,3-dipropylxanthine (DPCPX) and the 1,3-dipropyl-8-phenylxanthine amine congener (XAC) were obtained by elongation of the aliphatic substituents in the 1 and 3 position and additional changes of the substituents in the 8 position.

$[^3H]$DPX proved to be a potent, A_1-specific ligand in bovine brain but had 50- to 1000-fold lower affinities in brain membranes of other species. Other disadvantages are high levels of nonspecific binding and a relatively low specific radioactivity. All these properties were remarkably improved by the recently introduced radioligands $[^3H]$XAC (Jacobson et al. 1986) and $[^3H]$DPCPX (Bruns et al.

Table 4. Antagonist radioligands for A_1 receptors

Radioligand	Rat brain K_D (nM)	Nonspecific binding at K_D	Specific activity Ci/mmol
$[^3H]$DPX	68	40 %	13
$[^3H]$XAC	1.2	20 %	103
$[^3H]$DPCPX	0.18	1.3 %	105

1987a). They had 50- and 380-fold higher affinities for A_1 receptors of rat brain, respectively. Important additional advantages are the low nonspecific binding of $[^3H]$DPCPX and the high specific radioactivity of both ligands (Table 4).

The structural specificity of the new antagonist radioligands in rat brain membranes are consonant with an interaction at A_1 receptors which is identical to that obtained by agonist radioligands such as $[^3H]$PIA. An important additional property of DPCPX is its remarkably high selectivity for the A_1 adenosine receptor (Table 5). Theophylline, DPX and XAC are only 3- to 8-fold more potent antagonists at the A_1 receptor of rat brain compared to the A_2 receptor of rat striatum as determined by $[^3H]$NECA binding. In contrast, DPCPX has a 700- to 1100-fold A_1-selectivity over the A_2 receptor (Lee and Reddington 1986, Bruns et al. 1987a, Lohse et al. 1987).

The high affinity of $[^3H]$DPCPX for A_1 adenosine receptors has also been observed in membrane preparations of

Table 5. Selectivity of adenosine receptor antagonists

Compound	A_1 Receptor rat brain $[^3H]$PIA	A_2 Receptor rat striatum $[^3H]$NECA	A_1 Selectivity A_2/A_1
	K_i (nM)		
Theophylline	11,000	32,000	2.9
DPX	70	550	7.9
XAC	3.5	24	6.9
DPCPX	0.3	340	1,130

several other tissues. Saturation experiments showed low K_D-values ranging from 0.05 nM for bovine brain to 0.49 nM for rat heart membranes (Lohse et al. 1987). Nonspecific binding was negligible in brain and fat cell membranes and reasonably low (15% at K_D) in heart membranes. The binding capacity varied between 580 and 1,460 fmol per mg protein for rat fat cell and bovine brain preparations and very low values of 20 - 60 fmol per mg protein in rat and bovine cardiac tissues. These data are in good agreement with those obtained previously for tritiated and iodinated ligands.

Radioligand binding in intact cells

The excellent binding properties of [^{3}H]DPCPX have encouraged us to use this ligand for the analysis of A_1 receptors in intact cells. Previous studies with various agonist ligands did not provide satisfactory results due to high levels of nonspecific binding, a general problem of binding studies in intact cells (Lohse et al. 1986). By using [^{3}H]DPCPX, binding to A_1 receptors can readily be demonstrated in intact A_1-selective cells such as cardiomyocytes and fat cells.

We have first studied rat ventricular myocytes which have a very low receptor density. The binding was saturable with 68 fmol per 10^6 cells, which is equivalent to a binding capacity of 40 000 sites per cell. This receptor density is considerably lower than that reported for muscarinic and ß-adrenergic receptors on intact cardiomyocytes. The K_D-value is 0.48 nM and thus close to that determined in membrane preparations.

[^{3}H]DPCPX binding was further characterized in competition experiments with several adenosine agonists and antagonists. All competition curves were monophasic with slope factors not different from unity. DPCPX was the most potent compound with a K_i-value of 0.46 nM followed by the two agonists R-PIA and NECA and the antagonist theophylline.

Different results for agonists were obtained in membrane preparations of ventricular myocytes. The competition curves for agonists such as R-PIA were shallow and were significantly better fitted by a two-site model with 64% of the sites in the high affinity state (K_H 1.2 nM) and the remainder in the low affinity state (K_L 60 nM). These results show that two affinity states of A_1 receptors can be distinguished by agonists. In the presence of 100 µM GTP the competition curve for R-PIA was shifted to the right, and most of the sites were converted to the low affinity state. Under this condition, the competition curve obtained with membranes is fairly close to that observed with intact myocytes. Obviously, all of the adenosine receptors in the intact cell are in a low affinity state for the agonist, as would be expected from the high intracellular concentration of GTP.

In order to compare radioligand binding to intact cells with the subsequent cellular response to the A_1 receptor stimulation, we measured the inhibitory effect on isoprenaline-induced cAMP accumulation in the same cells. Table 7 shows the IC_{50}-values for the inhibition of cAMP accumulation and the inhibition constants for [^{3}H]DPCPX binding in intact cells and membrane preparations.

Table 7. Pharmacological profile of A_1 adenosine receptors in rat ventricular myocytes

Compound	cAMP accumulation	[^{3}H]DPCPX binding cells	membranes	
Agonists	IC_{50} (nM)	K_i (nM)	K_H (nM)	K_L (nM)
R-PIA	60	50	1.2	60
NECA	360	440	4.3	350
S-PIA	16,000	4,300	66	3,450
Antagonists	K_B (nM)	K_i (nM)	K_i (nM)	
DPCPX	0.15	0.46	0.42	
XAC	20	34	10.3	
Theophylline	12,000	4,900	2,900	

It can be seen that the IC_{50}-values of agonists correlate well with the corresponding affinities for [^{3}H]DPCPX binding sites in intact cells and the low affinity state in membranes. Obviously, the cellular effects of A_1 agonists are mediated by the receptor in the low affinity state. The high affinity state for agonists is only detectable in membranes and does not appear to be relevant for the effective potency of agonists.

On the other hand, the binding of antagonists to myocardial membranes showed only one state of homogeous affinity. Therefore, a close correlation between the effects on cAMP levels in intact cells and the inhibition of radioligand binding in cells and membranes was observed. These data suggest that spare receptors are not involved in the transduction of the receptor signal to the intracellular response in cardiomyocytes.

Photoaffinity ligands

Several photoaffinity probes have been synthetized and used for the biochemical characterization of A_1 adenosine receptors. Most of the compounds are based on the structure of the N^6-substituted adenosine analogs and are available in the radioiodinated form. $[^{125}I]$R-2-Azido-N^6-p-hydroxyphenylisopropyladenosine ($[^{125}I]$AHPIA) is a photolabile derivative of R-HPIA in which an azido group has been substituted at the 2 position of the purine ring (Klotz et al. 1985). $[^{125}I]N^6$(4-azido-3-iodobenzyl)adenosine ($[^{125}I]$AzBA) and $[^{125}I]N^6$-2-(4-azido-3-iodophenyl)ethyladenosine ($[^{125}I]$AZPNEA) are both derived from the parent compounds ABA and APNEA and each possess a functional 4-azido group at the benzyl or phenyl ring respectively (Choca et al. 1985, Stiles et al. 1986). All of the photoaffinity ligands have an apparently high affinity for the A_1 receptor and show K_D-values of approximately 1 nM.

Photoaffinity labeling with these compounds has identified the ligand-binding subunit of the A_1 receptor in several tissues as a peptide with M_r-values of 35,000 - 38,000. This peptide has all of the pharmacological properties of the A_1 receptor. In addition it has been demonstrated that the A_1 receptor of brain and fat cells of several species are glycoproteins with a common core protein of M_r 32,000 which is differently glycosylated (Stiles 1986, Klotz and Lohse 1986).

RADIOLIGANDS FOR A_2 ADENOSINE RECEPTORS

Radiolabeled agonists

5'-N-ethylcarboxamidoadenosine (NECA) and 5'-N-cyclopropylcarboxamidoadenosine (CPCA) are the most potent agonists for A_2 adenosine receptors which mediate a stimulation of adenylate cyclase activity. The first attempts to use tritiated CPCA and NECA in membrane preparations of rat brain cortex, rat liver and human platelets have been largely unsuccessful (Daly et al 1979, Schütz et al. 1982, Hüttemann et al. 1984). The results were not consistent with known properties of A_2 receptors since radioligand binding could not be inhibited by N^6-substituted adenosine analogs and most of the adenosine antagonists had only low affinities. Similar observations were reported for the binding of $[^3H]$2-chloroadenosine in human placenta membranes (Fox and Kurpis 1983).

A considerable improvement was achieved when binding of

[^{3}H]NECA in rat striatal membranes to both A_1 and A_2 receptors was found. The A_1 component of the binding could be eliminated by pretreatment with the sulfhydryl alkylating agent N-ethylmaleimide (Yeung and Green 1984) and more specifically with the A_1-selective agonist CPA (Bruns et al. 1986). In the presence of 50 nM CPA 98% of the A_1 receptor binding but only 7% of the A_2 receptor binding of [^{3}H]NECA is displaced. Under these conditions NECA bound with high affinity to two components of the A_2 receptor (3.5 and 35 nM) and exhibited a structure-activity profile which is consonant with an interaction at A_2 receptors (Bruns et al. 1986). NECA was the most potent agonist followed by 2-chloroadenosine and 2-phenylaminoadenosine. In agreement with adenylate cyclase studies NECA was 12-fold more potent than R-PIA as competitor of [^{3}H]NECA binding.

[^{3}H]NECA has also been employed for a successful characterization of A_2 receptors in the PC12 pheochromocytoma cell line (Williams et al. 1987). The radioligand bound with high affinity to a single class of binding sites in adenosine deaminase treated membranes. The binding was unaffected by the presence of low concentrations of CPA (50 nM) or $MgCl_2$ but was significantly decreased by guanylylimidodiphosphate suggesting the coupling of the A_2 receptor to a G-protein. In addition, the rank order of activity of agonists and antagonists was consistent with the selective labeling of an A_2 receptor in PC12 cells.

[^{3}H]NECA binding in human platelets

In contrast to the above mentioned studies with striatal membranes and PC12 cells [^{3}H]NECA binding to human platelet membranes has been hampered by the interference of a large amount of non-receptor binding sites (Hüttemann et al. 1984). We have therefore attempted to separate the A_2 receptor of human platelets from the non-receptor component by solubilization and chromatographic separation.

Human platelet membranes were solubilized with 1% CHAPS and chromatographed on Sepharose CL-6B. The typical elution profile yielded two [^{3}H]NECA binding peaks. The first peak was eluted with the void volume of the column together with the major protein peakand contained 10 to 25% of the total [^{3}H]NECA binding. The second peak was clearly separated and contained the majority of the [^{3}H]NECA binding. If nonspecific binding was defined with 100 μM 2-chloroadenosine, [^{3}H]NECA was displaced in both peaks. In contrast, 100 μM R-PIA displaced [^{3}H]NECA only from the first peak.

Further characterization of the two peaks showed that

Table 8. Pharmacological profile of A_2 receptors in human platelets

Compound	Adenylate cyclase	[^{3}H]NECA binding peak I	peak II	membranes
	EC_{50} (nM)	K_i (nM)		
Agonists				
NECA	190	70	330	500
PAA	350	130	8,800	6,200
R-PIA	1,300	1,700	>100,000	>100,000
ClA	1,900	1,000	4,100	6,300
CPA	3,000	3,400	45,000	>100,000
S-PIA	19,000	10,000	>100,000	>100,000
Antagonists	K_i (nM)			
XAC	24	55	>100,000	>100,000
DPX	210	180	>100,000	>100,000
Theophylline	13,000	20,000	>100,000	>100,000

[^{3}H]NECA binding to the first peak was saturable with a K_D of 42 nM and a maximal binding capacity of 505 fmol per mg protein. Nonspecific binding was 35% at K_D and 20% under usual assay conditions. A saturable binding was also obtained for peak II with a K_D of 234 nM and a very high binding capacity of 89,000 fmol per mg protein.

The pharmacological profile of both peaks was examined with several agonists and antagonists and compared with the EC_{50} values for stimulation of adenylate cyclase activity in platelet membranes (Table 8).

The K_i-values for the [^{3}H]NECA binding of the first peak agree reasonably well with the EC_{50} values of the cyclase experiments. This indicates that the first peak contains the A_2 receptor of the platelet. On the other hand, [^{3}H]NECA binding to the second peak was effectively inhibited only by NECA itself and 2-chloroadenosine whereas both isomers of PIA as well as all antagonists were ineffective in concentrations up to 100 µM. The structure-activity profile of this peak resembles very much the binding of [^{3}H]NECA to platelet membranes.

We believe therefore that we have separated a large non-receptor component of [^{3}H]NECA binding which interferes with the analysis of A_2 adenosine receptors in platelet membranes. At present, the properties and the possible functional role of the additional NECA-binding sites are unknown. Our results show that gel filtration provides a useful procedure to separate the A_2 receptor from the non-receptor binding sites and to perform reliable binding studies of the solubilized A_2 receptor of human platelets.

Radiolabeled antagonists

Two recently developed antagonist radioligands are now available for labeling A_2 adenosine receptors. These are the [^{3}H]xanthine amine congener of 1,3-dipropyl-8-phenyl-xanthine ([^{3}H]XAC) and [^{3}H]-N-[2-(dimethylamino)ethyl]-N-methyl-4-(1,3-dipropylxanthine-8-yl)benzenesulfonamide ([^{3}H]PD 115,199). Both compounds are not selective for A_2 adenosine receptors and, therefore, can only be used in subtype selective cells or in the presence of A_1 receptor ligands to eliminate the A_1 component of binding.

[^{3}H]XAC appears to bind with sufficient affinity (K_D 12 nM) to A_2 receptors of human platelet membranes but is associated with a high proportion (75%) of nonspecific binding at K_D (Ukena et al. 1986). The pharmacological profile of [^{3}H]XAC binding sites was similar to that obtained in adenylate cyclase studies (Table 9).

[^{3}H]PD 115,199 was used to identify A_2 receptors in rat striatal membranes (Bruns et al. 1987b). In the presence of 20 nM DPCPX a K_D of 2.6 nM was obtained. Nonspecific bind-

Table 9. Pharmacological profile of A_2 adenosine receptors

Compound	Rat striatum		Human platelets
	[^{3}H]NECA[a]	[^{3}H]PD 115,199[b]	[^{3}H]XAC[c]
		K_i (nM)	
Agonists			
NECA	10	49	330
PAA	119	470	
R-PIA	124	980	5,700
Antagonists			
PD 115,199	14	4.1	
XAC			17
DPX	863	370	190
8-PT	848		1,700
Theophylline	25,300		29,000

[a]Bruns et al. (1986), [b]Bruns et al. (1987b), [c]Ukena et al. (1986)

ing was approximately 21% of total binding. The pharmacological profile of the [^{3}H]PD 115,199 sites was consistent with a binding to a high affinity A_2 receptor. PD 115,199 and XAC were the most potent antagonists, whereas theophylline was 1,000 to 10,000 times less potent. Compared to [^{3}H]NECA binding, agonists were consistently about 5-fold less potent to displace [^{3}H]PD 115,199, while antagonists were about 3-fold more potent. Binding of the new antagonist ligand to the low affinity A_{2b} receptors could not be detected.

CONCLUSIONS

A large number of radioligands for the analysis of adenosine receptor subtypes has been developed and successfully used. The most suitable radioligand for labeling A_1 receptors appears to be the antagonist ligand [^{3}H]DPCPX. This compound exhibits a high subtype selectivity, a low level of nonspecific binding and a high specific radioactivity. These properties have enabled the first satisfactory binding studies of A_1 receptors in intact cells.

A_1 receptors can also be labeled by several agonist ligands such as [^{3}H]PIA, [^{3}H]CHA and [^{3}H]CPA. The latter would probably be preferable because of its twofold higher affinity than R-PIA, but experience with this new ligand is still limited. Iodinated ligands such as [^{125}I]HPIA provide a very high specific radioactivity and may offer advantages for the identification of A_1 receptors in tissues with low receptor densities. Several iodinated photoaffinity ligands such as [^{125}I]AHPIA and [^{125}I]AzBA have been synthetized and used for covalent labeling of the A_1 receptor.

Less radioligands for A_2 adenosine receptors are available, but the possibilities to study this receptor subtype have been considerably improved. [^{3}H]NECA can now be used as a radioligand in several tissues such as striatum, pheochromocytoma cells and human platelets with the precautions mentioned. [^{3}H]PD 115,199 appears to be a promising antagonist ligand for the analysis of striatal A_2 receptors.

ACKNOWLEDGEMENT

The author's research cited herein has been supported by the Deutsche Forschungsgemeinschaft (Schw 83/13-3, SFB 320).

REFERENCES

Böhm, M., Brückner, R., Neumann, J., Schmitz, W., Scholz, H., Starbatty, J., 1986, Naunyn-Schmiedeberg's Archives of Pharmacology, 332:403-405.

Bruns, R.F., Daly, J.W., Snyder, S.H., 1980, Proceedings National Academy of Sciences USA,77:5547-5551.

Bruns, R.F., Fergus, J.H., Badger, E.W., Bristol, J.A., Santay, L.A., Hartman, J.D., Hays, S.J., Huang, C.C., 1987a, Naunyn-Schmiedeberg's Archives of Pharmacology, 335:59-63.

Bruns, R.F., Fergus, J.H., Badger, E.W., Bristol, J.A., Santay, L.A., Hays, S.J., 1987b, Naunyn-Schmiedeberg's Archives of Pharmacology, 335:64-69.

Bruns, R.F., Lu, G.H., Pugsley, T.A., 1986, Molecular Pharmacology, 29:331-346.

Choca, J.I., Kwatra, M.M., Hosey, M.M., Green, R.D., 1985, Biochemical and Biophysical Research Communications, 131:115-121.

Daly, J.W., Nimitkitpaisan, Y., Pons, F., Bruns, R.F., Smellie, F., Skolnick, P., 1979, Pharmacologist, 21:253.

Daly, J.W., Butts-Lamb, P., Padgett, W., 1983, Cellular and Molecular Neurobiology, 3:69-80.

Fox, I.H., Kurpis, L., 1983, The Journal of Biological Chemistry, 258:6952-6955.

Haubrich, D.R., Williams, M., Yarbrough, G.G., Wood, P.L., 1981, Canadian Journal of Physiology and Pharmacology, 59:897-1020.

Hüttemann, E., Ukena, D., Lenschow, V., Schwabe, U., 1984, Naunyn-Schmiedeberg's Archives of Pharmacology, 325:226-233.

Jacobson, K.A., Ukena, D., Kirk, K.L., Daly, J.W., 1986, Proceedings National Academy of Sciences USA, 83:4089-4093.

Klotz, K.-N., Cristalli, G., Grifantini, M., Vittori, S., Lohse, M.J., 1985, The Journal of Biological Chemistry, 260:14659-14664.

Klotz, K.-N., Lohse, M.J., 1986, Biochemical and Biophysical Research Communications, 140:406-413.

Lee, K.S., Reddington, M., 1986, Brain Research, 368:394-398.

Linden, J., Patel, A., Sadek, S., 1985, Circulation Research, 56:279-284.

Lohse, M.J., Klotz, K.-N., Lindenborn-Fotinos, J., Reddington, M., Schwabe, U., Olsson, R.A., 1987, Naunyn-Schmiedeberg's Archives of Pharmacology, 336:204-210.

Lohse, M.J., Klotz, K.-N., Schwabe, U., 1986, Molecular Pharmacology, 30:403-409.

Lohse, M.J., Lenschow, V., Schwabe, U., 1984, Molecular Pharmacology, 26:1-9.

Londos, C., Cooper, D.M.F., Wolff, J., 1980, Proceedings National Academy of Sciences USA, 77:2551-2554.

Sattin, A., Rall, T.W., 1970, Molecular Pharmacology, 6:13-23.

Schütz, W., Tuisl, E., Kraupp, O., 1982, Naunyn-Schmiedeberg's Archives of Pharmacology, 319:34-39.

Schwabe, U., 1985. In Methods in Pharmacology, Vol. 6, edited by D.M. Paton (Plenum Press, New York, London), p. 239-278.

Schwabe, U., Lenschow, V., Ukena, D., Ferry, D.R., Glossmann, H., 1982, Naunyn-Schmiedeberg's Archives of Pharmacology, 321:84-87.

Schwabe, U., Trost, T., 1980, Naunyn-Schmiedeberg's Archives of Pharmacology, 313:179-187.

Stiles, G.L., 1986, The Journal of Biological Chemistry, 261:10839-10843.

Stiles, G.L., Daly, D.T., Olsson, R.A., 1985, The Journal of Biological Chemistry, 260:10806-10811.

Stiles, G.L., Daly, D.T., Olsson, R.A., 1986, Journal of Neurochemistry, 47:1020-1025.

Ukena, D., Jacobson, K.A., Kirk, K.L., Daly, J.W., 1986, FEBS Letters, 199:269-274.

Van Calker, D., Müller, M., Hamprecht, B., 1978, Nature, 276:839-841.

Williams, M., Abreu, M., Jarvis, M.F., Noronha-Blob, L., 1987, Journal of Neurochemistry, 48:498-502.

Williams, M., Braunwalder, A., Erickson, T.J., 1986, Naunyn-Schmiedeberg's Archives of Pharmacology, 332:179-183.

Williams, M., Risley, E.A, 1980, Proceedings National Academy of Sciences USA, 77:6892-6896.

Yeung, S.-M.H., Green, R.D., 1984, Naunyn-Schmiedeberg's Archives of Pharmacology, 325:218-225.

THE SEARCH FOR PURINE AND RIBOSE-SUBSTITUTED ADENOSINE ANALOGS WITH POTENTIAL CLINICAL APPLICATION

James A. Bristol, Alexander J. Bridges, Robert F. Bruns, David A. Downs, Thomas G. Heffner, Walter H. Moos, Daniel F. Ortwine, Deedee L. Szotek, and Bharat K. Trivedi

Parke-Davis Pharmaceutical Research Division
Warner-Lambert Company
2800 Plymouth Road, Ann Arbor, Michigan 48105 USA

Adenosine is an endogenous purine, which may act as a local hormone via activation of subpopulations of extracellular receptors. The purinergic receptor system (Figure 1) was defined with adenosine A1 receptors mediating a reduction of c-AMP in many tissues and A2 receptors mediating an increase of c-AMP (Van Calker et al. 1979, Londos et al. 1980). Receptor binding techniques for A1 receptors, (Bruns et al. 1980) and A2 receptors (Bruns et al. 1986) greatly aided in the rapid characterization of a variety of agonists for these receptor subpopulations. Further subdivision of the A2 receptor into high affinity (A2a) and low affinity (A2b) subtypes was proposed (Daly et al. 1983 and Bruns et al. 1986).

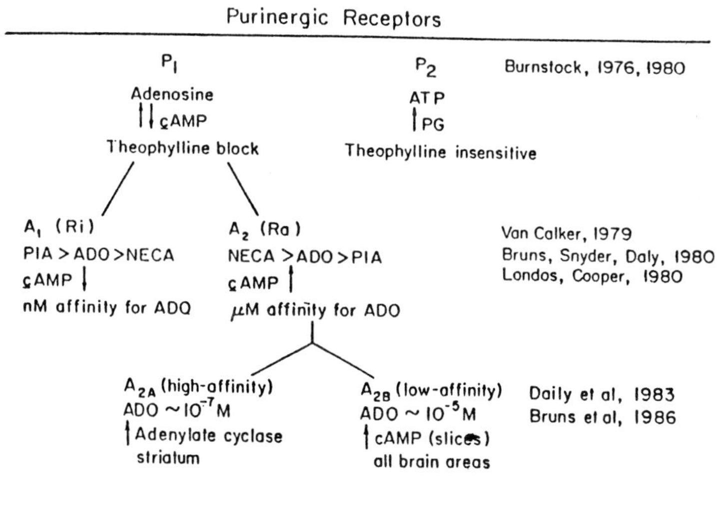

Figure 1. The Purinergic Receptor System.

Using receptor binding techniques, CHA, (R)-PIA, and NECA were characterized as high affinity A1 agonists (Bruns et al. 1980). With the development of an A2 receptor binding assay (Bruns et al. 1986). NECA was characterized as a high affinity A2a agonist. CV-1808, originally developed as an antianginal agent (Kawazoe, et al. 1980), was demonstrated to have high affinity and selectivity for A2a receptors. (Figure 2).

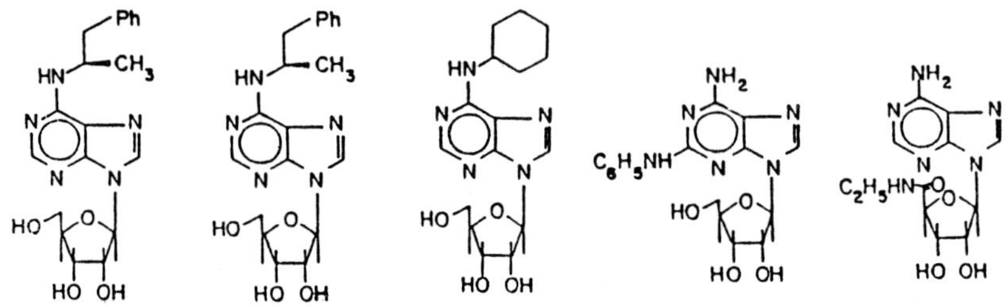

(R)-PIA	(S)-PIA	CHA	CV-1808	NECA
A_1 1.17nM	A_1 49.3nM	A_1 1.31	A_1 599nM	A_1 6.2nM
A_{2a} 124nM	A_{2a} 1814nM	A_{2a} 362	A_{2a} 116nM	A_{2a} 11.7nM

Figure 2. Receptor Binding Affinities (Ki) for Adenosine Agonists

Our interest in this emerging field began shortly after the initial reports of high affinity A1 binding (Bruns et al. 1980) and the demonstration of behavioral depressant effects of (R)-PIA (Snyder et al. 1981). It had been recognized as early as 1975 (Vapaatalo et al.) that many of the central and peripheral effects (inhibition of motor activity, analgesia, increased barbiturate sleeping time, hypothermia, hypotension, and negative chronotropic and inotropic effects) of (R)-PIA and the diastereomer (S)-PIA were stereospecific, with (S)-PIA always less potent.

Our early research on the behavioral effects of adenosine agonists characterized CHA, (R)-PIA, (S)-PIA, NECA, and CV-1808 in a motor activity and screen test (MAST) which measures in mice the inhibition of dark-stimulated motor activity and the ability to cling to an inverted screen, a measure of ataxia. Clinically useful antipsychotic agents cause ataxia at doses well above those which inhibit motor activity in this test, and similarly, certain adenosine agonists (Table 1) possess an antipsychotic-like profile.

Of these, the A2a selective agonist, CV-1808 clearly had the best antipsychotic-like profile (Heffner et al. 1985).

A more definitive test for antipsychotic-like activity is the continuous unsignaled (Sidman) avoidance responding test in rats. Evaluation of the same series of reference adenosine agonists in this test (Table 1) suggests again that certain of the adenosine agonists, (S)-PIA and particularly CV-1808, possess an antipsychotic-like profile in this model (Heffner, et al. 1985).

Table 1. Effects of Reference Adenosine Agonists in the Motor Activity (MA) and Ataxia (ATX) Test (MAST) and in Sidman Avoidance.

Compound	Mouse MAST ED50 mg/kg, ip		Rat ED50 mg/kg po
	MA	ATX	Sidman
(R)-PIA	0.8	2.0	0.36 (*)
(S)-PIA	15.6	75	14.7
NECA	0.05	0.22	1.76 (*)
CHA	0.22	2.0	0.56
CV-1808	1.75	>100	2.87

(*) non-selective for avoidance vs. escape

Although this is a small series of compounds, the results in the MAST and Sidman avoidance tests suggested to us that an adenosine agonist with high affinity for adenosine A2a receptors would be a novel and potentially interesting strategy for the identification of a non-dopaminergic antipsychotic agent. Accordingly, we began an integrated program designed to synthesize and evaluate novel adenosine agonists with high affinity for adenosine A2a receptors. A secondary objective was to identify agents with both high affinity and high selectivity for the A2a receptor. This strategy was bolstered by our finding that there is a relatively high regional distribution of A2a receptors in the striatum, nucleus accumbens, and olfactory tubercle, (Reddington et al. 1986), areas implicated in antipsychotic drug action.

Early in the program, we evaluated a homologous series of N-6 cycloalkyladenosines (Table 2) (Moos et al. 1985). This led to some very potent and highly A1-selective adenosines, with N-6 cyclopentyladenosine (CPA) identified as the adenosine agonist with the highest affinity known for A1 receptors (Ki = 0.59nM); however, none of the N-6 cycloalkyladenosines demonstrated improved A2a receptor affinity.

Table 2. Affinities of N-6 Cycloalkyl Adenosines in A1 and A2a Receptor Binding. (Moos, et al., 1985)

Compound	A_1 Ki (nM)	A_{2a} Ki (nM)
N-6 Cyclopropylado	3.19	1240
N-6 Cyclobutylado	0.77	263
N-6 Cyclopentylado	0.59	462
N-6 Cyclohexylado	1.31	363
N-6 Cycloheptylado	2.79	1690
N-6 Cyclooctylado	4.03	2450
N-6 Cyclodecylado	100	5190

In the course of our work, we also evaluated a series of ribose-modified (R)-PIA analogs and in virtually every case found that deviation from the ribose moiety produced a strikingly detrimental effect on receptor affinity (Taylor et al. 1984).

An alternative strategy was to evaluate certain C-2 and 5'- modifications of potent N-6 substituted adenosines. To accomplish this, we synthesized a series of analogs of CPA and (R)-PIA with substituents at the C-2 and 5'- positions (Table 3). Inspection of the data in Table 3 indicates that these two modifications can produce varying results depending upon the nature of the substituent. For example, the addition of an amino group at C-2 reduces A1 affinity five to ten fold, but also affords a greater reduction in A2a affinity, leading to more A1 selective agents. By contrast, substitution by phenylamino at C-2 to produce 2-(phenylamino)CPA or 2-(phenylamino)PIA (i.e., N-6 substituted analogs of CV-1808) results in a dramatic reduction in A1 affinity (20-fold for CPA and 200-fold for PIA) with relatively minimal effects on A2a affinity; however, the absolute A2a receptor affinities of these compounds are still rather weak.

In a similar fashion, 5'-modification can produce more A1-selective agents when the 5'-substituent is chloro or hydrogen, but when the 5'-hydroxymethyl is oxidized to an acid and converted to an N-ethyl carboxamide, relatively more A2a selective compounds are formed; however, as demonstrated by Olsson et al. (1986), this modification results in agents retaining significant A1 affinity.

Table 3. Effects of 2- and 5'-Modification on A1 and A2a Receptor Binding of CPA and (R)-PIA

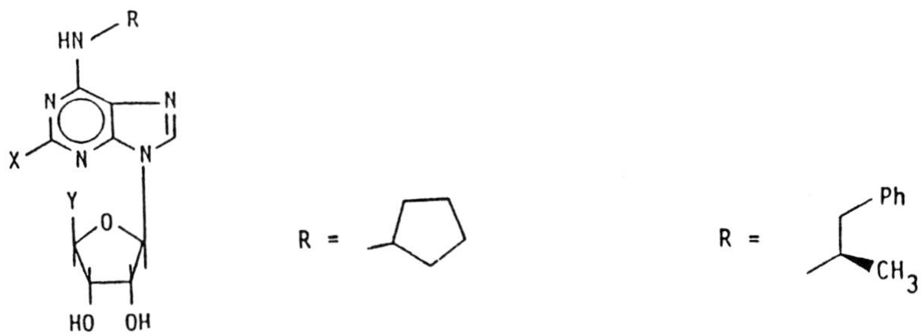

X	Y	KI (nM) A_1	A_{2a}	A_{2a}/A_1	Ki (nM) A_1	A_{2a}	A_{2a}/A_1
H	CH_2OH	0.59	461	782	1.17	124	106
NH_2	CH_2OH	2.89	6092	2108	9.6	1530	159
PhNH	CH_2OH	12.4	143	11.6	152	317	2.09
H	CH_2Cl	0.71	1866	2629	1.62	1078	665
H	CH_3	9.12	11592	1271	24.8	3086	124
H	CONHEt	0.50	133	267	1.36	46.2	34

Thus it became apparent that the N-6 substituent played a dominant role in controlling the absolute A1 affinity and we initiated a strategy of rational modification of this substituent as a means to produce compounds with high A2a affinity. Inspection of a homologous series of N-6 aralkyl adenosines (Table 4) clearly substantiated the importance of chirality at the alpha position for determining A1 affinity. Thus (R)-alpha- phenethyl adenosine and (R)-PIA are much more A1 selective and have higher absolute A1 affinities relative to their (S)-stereoisomers. Moreover, both (R)- and (S)- alpha phenethyl- and beta phenethyladenosines were more A1 selective than their unsubstituted analogs, demonstrating the importance of the alpha position for controlling A1 receptor affinity.

As predicted from our earlier behavioral results, A1 receptor selective adenosine agonists did not produce a selective profile in MAST whereas less A1 selective agents (N-6 benzyl, N-6 phenethyl, and (S)-PIA) were somewhat selective although relatively weak in MAST. These results suggested that an N-6-substituted adenosine with higher A2a affinity might have an improved profile in MAST with the potential for antipsychotic activity.

Table 4. Effects of N-6 Aralkyl Substitution on Receptor Binding, Motor Activity and Ataxia (MAST).

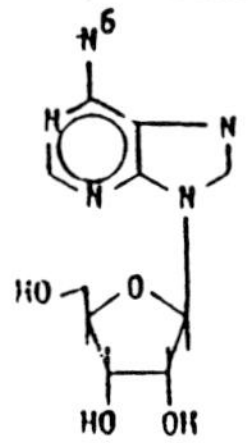

N^6	Receptor Binding Ki (nM) A_1	A_{2a}	Ratio A_{2a}/A_1	MAST mg/kg ED_{50} (IP) LA	SF
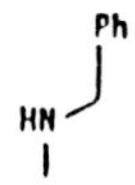	119	284	2.39	22	>100
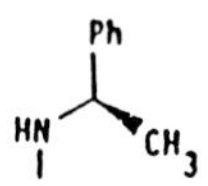	5.48	455	83.1	0.8	3.0
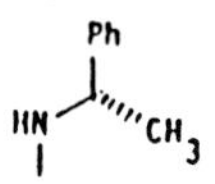	189	6337	33.5	>100	>100
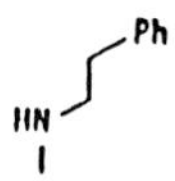	12.5	160	12.8	20	>100
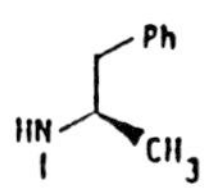	1.17	124	106	0.8	2.0
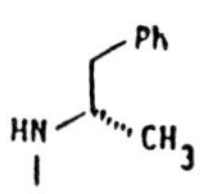	49.3	1814	36.8	15.6	75

Since alpha substitution seemed to control A1 receptor affinity, we began a systematic investigation of the effects of substitution at the beta position of N-6 2-phenethyladenosine. One of the first compounds to be synthesized was N-(2,2-diphenylethyl)adenosine (CI-936). This is the first N-6 substituted adenosine agonist identified with high A1 (Ki = 6.9 nM) and A2a (Ki = 24.7 nM) receptor affinity. As predicted from our earlier behavioral studies, CI-936 had an excellent separation in MAST, suggestive of an antipsychotic-like profile (Figure 3 and Table 5).

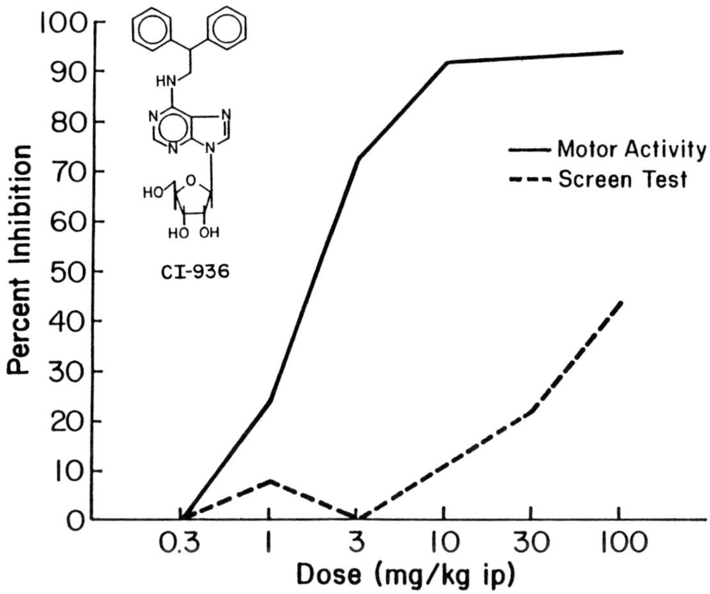

Figure 3. Dose-Response Relationship of Effects of CI-936 on Motor Activity and Screen Test Failure (Ataxia) in Mice

Moreover, in rats dosed orally, CI-936 showed an excellent profile of activity when evaluated for motor activity and ataxia, and in the Sidman avoidance test when compared with other useful antipsychotic agents (Table 5) (Bridges et al. 1987).

The identification of CI-936 as the first adenosine agonist substituted at N-6 with high affinity for A2a adenosine receptors, and with a very favorable profile of antipsychotic-like activity in animal models prompted us to thoroughly explore the structure-activity relationships of this interesting new series. The synthesis of a series of

Table 5. Comparison of CI-936 with Haloperidol and Thioridazine in the Motor Activity (MA) and Ataxia (ATX) Test and in Sidman Conditioned Avoidance.

Compound	Mouse ED50 mg/kg, ip		Rat ED50 mg/kg, po		Rat ED50 mg/kg,po
	MA	ATX	MA	ATX	Sidman
CI-936	1.3	145	2.2	31	2.5
Haloperidol	0.27	12	0.24	>1	0.3
Thioridazine	3.3	13	13.3	>30	19.7

higher and lower homologs of CI-936 (Table 6) (Bridges et al. 1987) quickly established that a two carbon chain was optimal and that further substitution at the beta carbon with small substituents (methyl and hydroxy) produced compounds that retained affinity. Of significance is that the 2-methyl analog of CI-936 is the first N-6 substituted adenosine with greater affinity for the A2a receptor than the A1 receptor.

Having established that a two carbon chain was optimal, we next investigated several cyclic analogs of CI-936 (Trivedi et al. 1987) (Table 7). Only the fluorenyl analog retained affinity, (Ki = 5 nM). suggesting a highly defined spatial relationship at N-6 which allowed the fluorenyl, but not other tricyclic ring systems to fit the receptor.

Subsequently, a thorough analysis of the effects of aryl ring substitution within the CI-936 series was undertaken. Computer assisted molecular modeling (Ortwine et al. Parke-Davis Pharmaceutical Division, unpublished results) established primary and accessory N-6 aryl binding regions for the A2a receptor. This led to the synthesis of a series of substituted diarylethyladenosines, several of which are shown in Table 8. As can be seen from this limited series, judicious placement of one or two chlorine substituents has a dramatic effect on both absolute affinity and receptor selectivity. One para chlorine substituent produces a slight diminution of affinity, while two para substituents produces a 20 fold reduction in affinity for both receptors. One or two meta chlorine substituents increase affinity, and 3,5-disubstitution on one ring produces a compound which is A2a receptor selective. The complete SAR and pharmacology of these A2a receptor selective agonists will be reported elsewhere.

Table 6. Effects of Chain Length and Substitution on A1 and A2a Receptor Binding

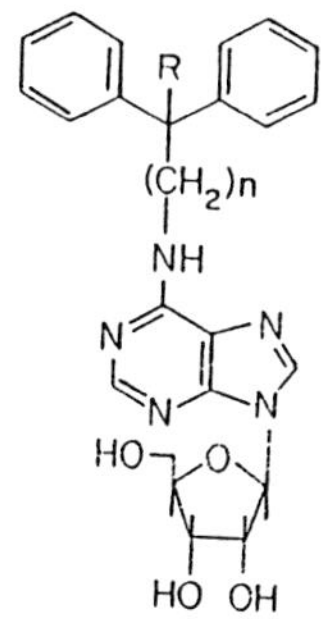

n	R	Ki (nM) A_1	Ki (nM) A_{2a}
1	H	6.86	24.6
0	H	479	3808
2	H	242	289
3	H	290	4653
4	H	85.1	2224
1	CH_3	37.4	29.7
1	C_3H_7	53.2	782
1	Ph	$>10^{-5}$	$>10^{-5}$
1	OH	6.8	50.7

Table 7. Cyclic Analogs of CI-936

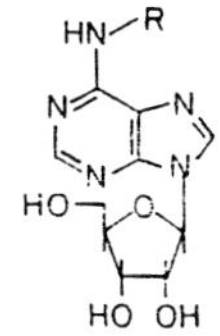

R	Ki (nM) A_1	Ki (nM) A_{2a}
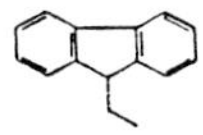	5.1	4.9
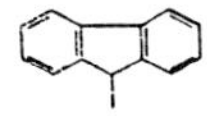	6200	42000
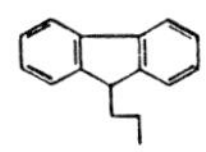	320	4300
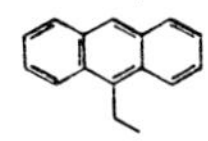	9000	29000
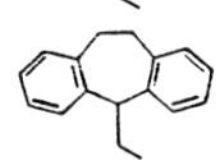	160	390

Table 8. Effects of Ring Substitution on Receptor Affinity and Selectivity of CI-936 Analogs.

	X	X'	Ki (nM) A_1	Ki (nM) A_{2a}	A_{2a}/A_1
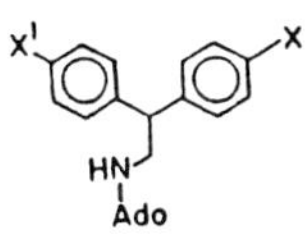	H	H	6.86	24.6	3.58
	Cl	H	15.3	60.7	3.96
	Cl	Cl	290	480	1.66
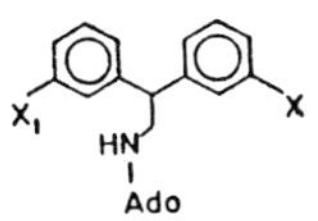	Cl	H	1.7	5.52	3.25
	Cl	Cl	2.8	4.0	1.4
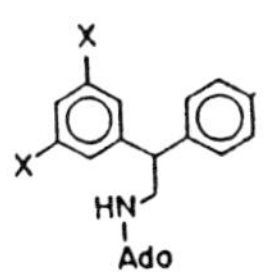	Cl	-	32.5	9.4	0.29

SUMMARY

One strategy for modern drug discovery is to identify agents which selectively modulate (either stimulate or antagonize) a receptor subpopulation, thus eliciting a desired pharmacological efffct (e.g. blockade of histamine H2 receptors). With the identification of subpopulations of adenosine receptors, the possibility exists for the development of novel therapeutic agents interacting with these receptors. We have identified a potent and selective series of adenosine agonists which have potential as a novel class of non-dopaminergic antipsychotic agents.

REFERENCES

Bridges, A. J., Moos, W. H., Szotek, D. L., Trivedi, B. K., Bristol, J. A., Heffner, T. G., Bruns, R. F., & Downs, D. A., 1987, Journal of Medicinal Chemistry, 30, in press.

Bruns, R. F., Daly, J. W., & Snyder, S. H., 1980, Proceedings of the National Academy of Sciences USA, 77, 5547.

Bruns, R. F., Lu, G. H., & Pugsley, T. A., 1986, Molecular Pharmacology, 29, 331.

Daly, J. W., Butts-Lamb, P. & Padgett, W., 1983, Cellular and Molecular Neurology, 3, 69.

Heffner, T. G. , Downs, D. A., Bristol, J. A., Bruns, R. F., Harrigan, S. E., Moos, W. H., Sledge, K. S., & Wiley, J. N., 1985, Pharmacologist, 27, 293.

Kawazoe, K., Matsumoto, N., Tanabe, M., Fujiwara, S., Yanagimoto, M., Hirata, M. & Kikuchi, K., 1980, Arzneimittel-Forschung, 30 (II) 1083.

Londos, C., Cooper, D. M. F., & Wolff, J., 1980, Proceedings of the National Academy of Sciences USA, 77, 2551.

Moos, W. H., Szotek, D. S., & Bruns, R. F., 1985, Journal of Medicinal Chemistry, 28, 1383.

Olsson, R. A., Kushachi, S., Thompson, R. D., Ukena, D., Padgett, W., & Daly, J. W., 1986, Journal of Medicinal Chemistry, 29, 1683.

Reddington, M., Erfurth, A., and Lee, K. S., 1986, Brain Research, 399,232.

Snyder, S. H., Katims, J. J., Annau, Z., Bruns, R. F., & Daly, J. W., 1981, Proceedings of the National Academy of Sciences USA, 78, 3260.

Taylor, M. D., Moos, W. H., Hamilton, H. W., Szotek, D. S., Patt, W. C., Badger, E. W., Bristol. J. A., Bruns, R. F., Heffner, T. G., & Mertz, T. E., 1986, Journal of Medicinal Chemistry, 29, 346.

Trivedi, B. K., Bristol. J. A., Bruns, R. F., Haleen, S. J., and Steffen, R. P., 1987, 195th ACS National Meeting, MEDI 45.

Vapaatalo, H., Onken, D., Neuvonen, P. J., and Westermann, E., 1975, Arzneimittel-Forschung, 25, 407.

Van Calker, D. Muller, M. & Hamprecht, B., Journal of Neurochemistry, 33, 999.

A FUNCTIONALIZED CONGENER APPROACH TO AGONISTS AND ANTAGONISTS FOR ADENOSINE RECEPTORS

Kenneth A. Jacobson,* Kenneth L. Kirk,* and John W. Daly

Laboratories of Chemistry* and Bioorganic Chemistry
NIDDK, National Inst. of Health, Bethesda MD.

INTRODUCTION

Recent structure-activity studies within a number of classes of drugs (adrenergics: Jacobson et al 1983; purines: Jacobson et al 1985b,c; histamine: Khan et al 1986; opioid alkaloids: Portoghese 1987) have demonstrated that many drugs may be attached to well-defined "carrier" molecules and still retain the ability to bind to the cell surface receptor site and produce a biological effect. A particular strategy for the design of carrierbound drugs is the "functionalized congener" approach, by which new analogs are synthesized with the regiospecific inclusion of a chain (containing a chemical functional group such as an amine or carboxylic acid at the terminus). This chain (Fig. 1, symbolized by —<) is situated at a nonsensitive position on the pharmacophore in order not to interfere sterically at the binding site. This is shown schematically in Fig. 1b. Derivatives of such a functionalized congener are used to delineate how biological activity is affected using that site of attachment and to probe what structural features may be optimal in a more elongated chain. The resultant chemically functionalized drug congener then may be attached through the reactive group on the chain to various organic moieties, such as amines and peptides. One goal is to be able to attach at a distance from the pharmacophore bulky groups, which would not be tolerated when bonded directly (Fig. 1c). Surprisingly, the receptor binding affinity of these analogs is often greater than that of the parent drug and does not necessarily diminish as the molecular weight is increased. Moreover, there is in some cases an energetically favorable interaction, such as a hydrogen bond or a non-specific interaction, between the attached group

Fig. 1

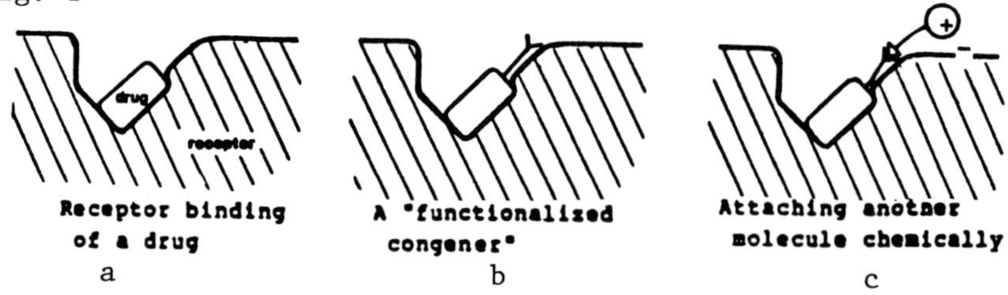

("carrier") and an accessory site on the receptor molecule or a nearby membrane component, which can increase the strength of binding of the conjugate (for example, the electrostatic interaction in Fig. 1c). Distal structural modifications then are used to optimize in the same derivative both potency and other desirable properties such as water solubility. A goal of such structural investigations is the eventual development of new therapeutic agents having favorably altered potency, selectivity (including targeted delivery), or duration _in vivo_. In addition, the functionalized congener approach is directed towards the synthesis of high affinity molecular probes for receptors.

Fig. 2. Structures of xanthine functionalized congeners.

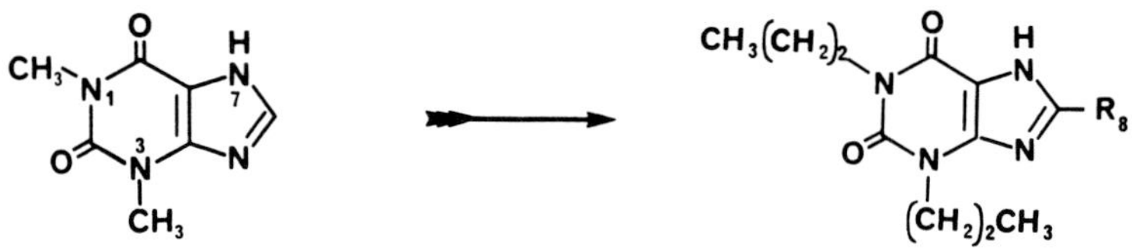

THEOPHYLLINE (ANTAGONIST)

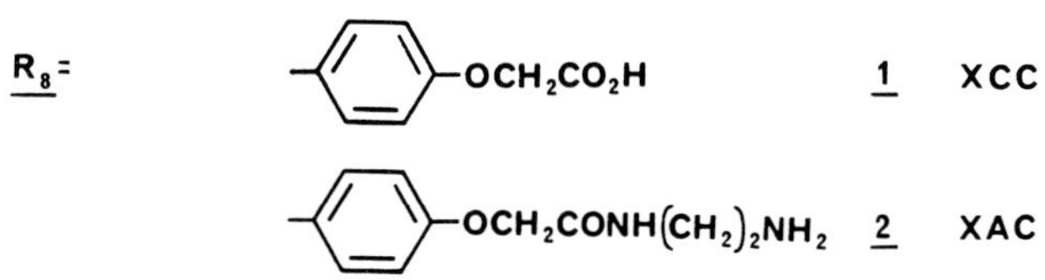

ADENOSINE AGONISTS AND ANTAGONISTS

Xanthine Functionalized Congeners

A series of functionalized congeners derived from 1,3-di-alkyl-8-arylxanthines, containing carboxylic or amino groups for coupling to carriers, was synthesized as adenosine antagonists (Jacobson et al 1985b). Based on the high receptor affinity associated with 8-phenyl substitutions (Bruns 1981; Bruns et al 1980), including the p-hydroxyphenyl analog, an 8-(p-carboxymethyloxy) substitutent was introduced to form a xanthine carboxylic acid congener (XCC, for $R_{1,3}$ = propyl, 1 in Fig. 2). Consistent with previous findings (Bruns et al 1983), replacement of the 1- and 3-position methyl groups with n-propyl groups increased the A_1-receptor affinity of the carboxylic acid congener and simple related amides (Jacobson et al 1985b).

Coupling of the carboxylic acid congener to diamines produced a series of amine derivatives, eg. 2 (Jacobson et al 1985b). The analog derived from ethylene diamine was prepared coveniently by the aminolysis of the ethyl ester of XCC. This 1,3-dipropylxanthine amine congener (XAC) initially was designed to be a synthetic intermediate in the synthesis of conjugates (for example by acylation or alkylation) but was, itself, of sufficiently high potency to warrant further *in vitro* and *in vivo* studies.

In initial screening in rat brain membranes, XCC (Ki for inhibition of N^6-cyclohexyladenosine binding was 58±3 nM) appeared to be non-selective between A_1- and A_2-receptor subtypes. However the rise in A_1-affinity in XAC, with a Ki-value in CHA binding of 1.2±0.5 nM, produced an apparent 40-fold selectivity for A_1-adenosine receptors compared to A_2-receptors (inhibition of 2-chloroadenosine-elicited cyclic AMP accumulation in brain slices). In a subsequent study of adenylate cyclase effects in platelet and pheochromocytoma cell membranes (Ukena et al 1986a), XCC was shown to be considerably less potent at these A_2-receptors. Thus, both XAC and XCC appear to be A_1-selective. Relatively high potency at adenosine receptors is associated with many other derivatives of XAC, particularly those containing distal amino groups, including those with additional alkyl substitution on the amine or longer spacer groups separating the amine from the primary pharmacophore. It must be stressed at this point that the degree of A_1-selectivity of 8-arylxanthines is species dependent. XAC binds to A_1-receptors 13-fold more weakly in guinea pig brain than in rat (Ukena et al 1986c).

The A_1 selectivity of the antagonist XAC is useful for investigation of receptors involved in the physiological functions of adenosine. XAC blocks peripheral actions of adenosine agonists, such as hypothermia, but when administered ip. in mice fails to block the locomotor depression induced by NECA or CHA (Seale et al 1987). In antagonizing the cardiovascular effects of NECA, XAC, at low doses (Table 1), reversed the cardiodepressant effects, but not the hypotensive effects (Jacobson et al 1985d; Fredholm et al

1987). The 20-fold cardioselectivity of this potent xanthine derivative suggested in vivo A_1-subtype selectivity. XCC was not subtype selective in the cardiovascular studies in vivo in rats. XAC also selectively blocked the cardiodepressant effects of 2-chloroadenosine and R-PIA as well as potentiation by adenosine of the positive chronotropic effect of nicotine (Evoniuk et al 1987).

In the isolated perfused kidney (Rossi et al 1987) XAC was roughly 3000-fold more potent than theophylline in reversing the vasoconstrictive effect of adenosine agonists in afferent arterioles (A_1-receptor mediated). The A_1-selectivity of XAC is apparent also in the kidney (20-fold), as indicated by the reversal of biphasic agonist effects on renin release (Churchill et al, unpublished).

Table 1. Potency of xanthine derivatives as adenosine receptor antagonists in vivo in rats. Results are expressed as the doses of the xanthine derivatives that produce a 2-fold decrease in the potency of N-5'-ethylcarboxamidoadenosine (NECA). Data taken from Fredholm et al (1987).

Drug	Heart rate mg/kg	Heart rate µmol/kg	Blood pressure mg/kg	Blood pressure µmol/kg	Ratio
Caffeine	3.39	17.5	3.63	18.7	1.07
Enprofylline	14.5	63.6	17.0	74.6	1.17
8-Phenyltheoph.	0.062	0.22	0.041	0.14	0.66
8-p-Sulphophenyl-theophylline	0.85	2.3	1.71	4.70	2.01
XCC	0.50	1.1	0.69	1.5	1.38
XAC	0.060	0.14	1.18	2.8	20

Various covalent conjugates of XCC and XAC illustrated the generality of derivatization through this chain structure. The pronounced effects on the potency of binding and on the receptor subtype selectivity produced by the presence of charged groups on the chain suggested the synthesis of a series of amino acid and peptide conjugates (Jacobson et al 1986a). In this series, the potency was not necessarily diminished as the size of the conjugate was increased. Conjugates in which the attached chain contained a free amino goup tended towards high A_1 potency, and in some cases A_1 selectivity.

Covalent conjugates of functionalized congeners and oligopeptides with peptide sequences resembling both natural hormones and unnatural synthetic sequences have been synthesized and found to bind to adenosine receptors. For example, XAC has been attached to the terminal pentapeptide of the putative neurotransmitter substance P (SP) resulting in a binary drug* (Table 2) capable of binding to either A_1 or SP receptors (Jacobson et al 1987b). We are now examining whether the activities of both components of such binary conjugates are discernable in vivo. Adenosine and SP agonists have hypotensive effects in common; a binary drug combining both their cardiovascular activities might thus have augmented physiological effects.

The "functionalized congener" approach provides a means for enhancing the water solubility of xanthine derivatives. For example, the D-lysyl conjugate of XAC (K_i-value in [^{3}H]CHA binding of 0.87±0.09 nM), had an aqueous solubility of 0.52 mM, compared to 3 μM for 1,3-dipropyl-8-(p-hydroxyphenyl)xanthine. In 0.1 M acetic acid the solubility of XAC is 5 mM, and that of ADAC (see below) is 12 mM.

Table 2. Potencies of the binary drugs and various truncated derivatives at adenosine A_1-receptors and at substance P (SP) receptors (Jacobson et al 1987b).

Peptide Derivative	K_i (nM) A_1	K_i (nM) SP
Substance P analogs		
H-Phe-Phe-Gly-Leu-Met-NH_2	>30,000	>10,000
H-Gln-Phe-Phe-Gly-Leu-Met-NH_2	-	450
Arg-Pro-Lys-Pro-Gln-Gln-Phe-Phe-Gly-Leu-Met-NH_2 (SP)	>30,000	1
Adenosine derivative		
N^6-ϕ-CH_2-CO-Phe-Phe-Gly-Leu-Met-NH_2	16±0.9	2,000
Xanthine derivatives		
H-Asp-XAC └Phe-Phe-Gly-NH_2	30±6.0	>10,000
H-Asp-XAC └Phe-Phe-Gly-Leu-Met-NH_2*	35±3.4	300

Adenosine Functionalized Congeners

N^6-Phenyladenosine is a highly potent inhibitor of binding of [^{3}H]CHA at high affinity A_1-adenosine receptors (roughly 6-fold greater affinity than adenosine, K_i = 3.2 nM). A site for elaboration of functionalized side chains was created at the para-position of N^6-phenyladenosine by introduction of a carboxymethyl substituent (Jacobson et al 1985c). A structural feature of certain analogs is the presence of functional groups, such as amino, carboxylic, and hydrazido groups, which can be linked to other molecules or to solid supports for affinity chromatography.

The carboxylic acid congener, 3, N^6-(p-carboxymethylphenyl)adenosine (Fig. 3), could be coupled to various alkyl and aryl amines using common condensing reagents, such as carbodiimides combined with 1-hydroxybenzotriazole as a catalyst, to give amide derivatives. It was found that analogs having aryl amides at this position tended to be more potent in A_1-receptor binding than alkyl amides. For this reason, an additional aryl group was included at this point leading to extended structures. Several peptide conjugates of 3, in which the site of attachment was a side chain amide-linked p-aminophenylalanine residue, were biologically active. Recently, compound 3 was linked to phosphatidylethanolamine by condensation in an attempt to target potentially drug-carrying liposomes to cells bearing adenosine receptors (Nye & Snyder 1986). Liposomes incorporating this adenosine lipid conjugate displaced [^{3}H]PIA in binding studies.

Fig. 3. Adenosine-derived functionalized congeners.

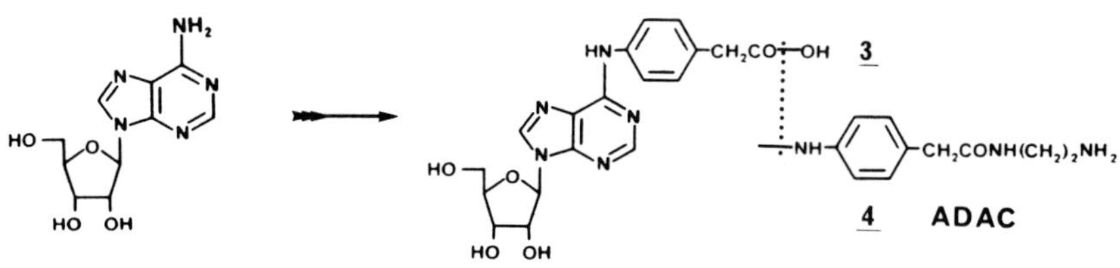

An adenosine amine congener (ADAC, 4, Fig. 3) derived from ethylenediamine, could be acylated readily using active esters, such as N-hydroxysuccinimide esters, to give amides. ADAC was an extremely potent member of the series, having a K_i-value of 0.85±0.35 nM at A_1 receptors. Binding of [^{3}H] ADAC showed K_D-values of 1.4 and 0.34 nM in rat and calf brain membranes, respectively (Jacobson et al 1987a). Thus, the affinity at A_1-receptor sites was improved by a factor of 5 over that of the parent N^6-phenyladenosine. It appeared that this enhancement was related to the presence of an amino group, presumably in the ammonium form predominant at physiological pH, since, after N-acetylation, the affinity was somewhat diminished. Furthermore, a hydrazide analog, which has a reduced pK_a of the terminal NH_2 and should not be protonated at physiological pH, has a significantly lower affinity. This pattern of a free amino group producing high affinity at A_1 sites parallels results for functionalized 1,3-dialkylxanthines as adenosine antagonists (see above). As depicted in Fig. 1c, this effect may be due to attraction of a charge on the drug to an anionic site on the receptor or to another anionic membrane component such as a phospholipid. This electrostatic explanation is also consistent with the low potency of anionic functionalized congeners.

To investigate the suitability of the amino group of ADAC as a position for generalized attachment, this amine congener was coupled to various prosthetic groups, which were chosen with specific applications in mind. One such group was p-hydroxyphenylpropionic acid, known for its application in augmenting radioiodination sites on proteins. The resulting amide adduct, bound to the A_1-adenosine receptor with a K_i-value in [^{3}H]CHA competition experiments on rat cerebral cortex membranes of 4.5 nM.

Fig. 4. Coronary vasoactivity of adenosine conjugates (molar potency ratio relative to adenosine = 1).

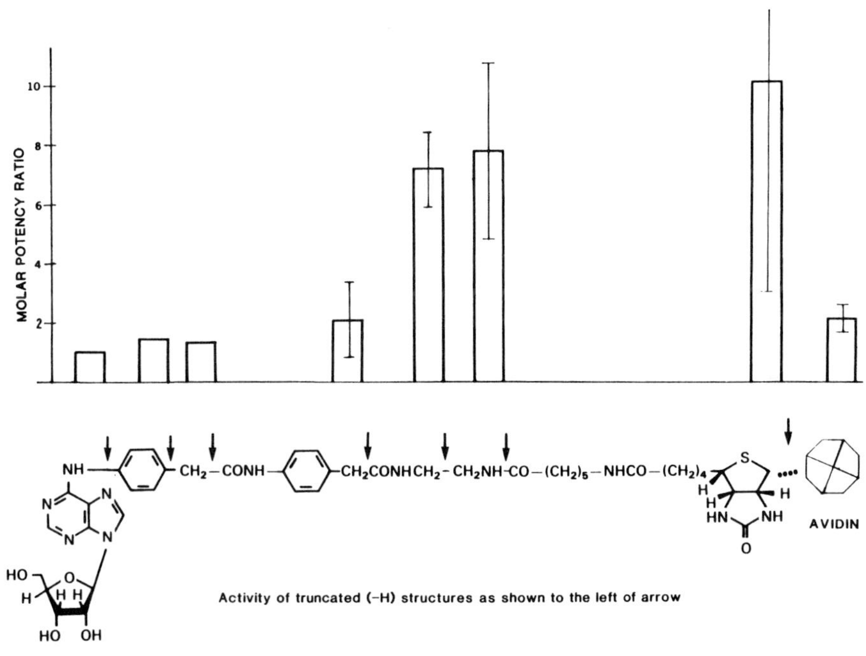

A number of adenosine derivatives designed through the functionalized congener approach were found to have enhanced

A_2-coronary vasoactivity, although not to the degree of NECA. In an in vivo study (Fig. 4, Jacobson et al 1986c), compounds bearing distal carboxamides, a biotin conjugate, and ADAC were considerably more potent than adenosine (estimated EC_{50} values for these analogues were between 120 and 240 nM, compared to 1200 nM for adenosine) in causing coronary vasodilation upon intracoronary infusion in anaesthetized, open-chest dogs. Although most N^6-substituted adenosines are highly selective for A_1 receptors, certain of the conjugates had much lower A_2/A_1 selectivity ratios when comparing A_2-coronary and A_1-brain adenosine receptors. The biotinyl-ε-aminocaproyl conjugate of ADAC had approximately the same A_2/A_1 selectivity ratio as 2-chloroadenosine, although chloroadenosine was nearly three fold more potent in these systems.

MOLECULAR PROBES FOR ADENOSINE RECEPTORS

As an approach to the detection and characterization of adenosine receptors, structural features which serve as markers for specific chemical or physical methods of measurement have been incorporated into receptor ligands, using the functionalized congener approach. In a broad category of molecular probes one may include radioactive markers, as well as spectroscopic and microscopic markers (such as fluorescent tags for the receptor, chemically- and photochemically-reactive affinity labels, and polymer-bound ligands (coupled to biopolymers or immobilized matrices).

A particularly successful approach to the first high affinity antagonist radioligand for adenosine receptors was based on the high potency of XAC at both A_1 and A_2 receptors. $[^3H]$XAC, with a specific activity of 157 C_i/mmol, was synthesized by catalytic reduction with tritium gas of the 1,3-diallyl analog (Jacobson et al 1986b). It bound specifically to A_1 receptors in brain membranes from various species, and gave receptor densities in close agreement with the values obtained using $[^3H]$PIA. K_D-values determined from Scatchard plots were 1.2 and 0.17 nM for rat and calf brain, respectively.

Quantitative autoradiography of brain slices using tritiated xanthine functionalized congeners has been carried out. Binding of $[^3H]$XCC (Jarvis et al 1987) was consistent with the occurence of A_1-receptors. The highest density of binding was observed in the molecular layer of the cerebellum and the CA-1 region of the hippocampus (95 and 76 fmol/mg tissue, respectively).

At A_2-receptors in human platelets (Ukena et al 1986b), $[^3H]$XAC was of sufficiently high affinity to permit measurement of specific saturable binding (K_D = 12 nM, B_{max}

= 1.1 pmol/mg protein). Moreover, displacement of [^{3}H]XAC yielded inhibition constants for various agonists and antagonists, which for the first time in an A_2-binding assay, corresponded closely to functional values for stimulation or for blockade of adenylate cyclase (Table 3).

Table 3. Pharmacological Profile of A_2 Adenosine Receptors of Human Platelets (data from Ukena et al 1986b).

Compound	[^{3}H]XAC binding K_i (µM)	Adenylate Cyclase EC_{50} or K_B (µM)
XAC	0.017	0.025
NECA	0.33	0.26
DPX	0.19	0.21
1,3-Dipropyl-8-phenyl-xanthine	1.0	2.1
8-Phenyltheophylline	1.7	1.9
XCC	1.8	2.4
8-(p-Sulfophenyl)-1,3-dipropylxanthine	2.4	1.9
R-PIA	5.7	3.1
8-(p-Sulfophenyl)theophylline	7.2	5.5
Theophylline	29	13.8
Caffeine	51	43

As an alternative to single atom radioisotopic replacement, we have the possibility of attaching to functionalized congeners prosthetic groups, such as metal complexing agents, designed to capture specifically various radioisotopes (Jacobson et al 1987a).

Attachment of a prosthetic group for radioiodination to the antagonist XAC was utilized to acheive covalent labelling of adenosine receptors (Stiles & Jacobson 1987). Previously, Stiles et al (1985) have reported the photoaffinity crosslinking of A_1-adenosine receptors. Their method uses an aryl amine-functionalized derivative of adenosine, N^6-(p-aminophenylethyl)adenosine (APNEA), which may be iodinated. The aryl amino group of the receptor-bound APNEA is accesible to the surrounding aqueous environment and reacts specifically with a commercial, bifunctional photo-crosslinking reagent N-succinimidyl-6-(4-azido-2-nitrophenylamino)hexanoate (SANPAH). After the formation of an amide bond with the bound APNEA, the azido group is photolyzed and the resulting nitrene forms an irreversible covalent bond with the receptor protein.

We have acylated XAC with the p-aminophenylacetyl (PAPA) group, both as a prosthetic group for the introduction of iodine and for photoaffinity crosslinking to the receptor protein (Stiles & Jacobson 1987, Figure 5). PAPA-XAC may be labelled on the aminophenyl ring with ^{125}iodine by a chloramine T method. [^{125}I]PAPA-XAC binds reversibly and specifically to A_1-adenosine receptors of bovine cerebral cortex with a K_d-value of 0.1 nM (unchanged in the presence of GTP).

In bovine brain membranes [^{125}I]PAPA-XAC may be crosslinked specifically to a polypeptide of MW 40,000 (Stiles & Jacobson 1987), similar to that determined previously for the A_1-receptor protein using agonist ligands. The receptor-bound ligand reacts chemically with the heterobifunctional crosslinking reagent SANPAH at a conc. of 38 µM. Irradiation at 290 nm results in irreversible crosslinking of the ^{125}iodine label to the receptor, as detected by autoradiographic gel electrophoresis. Covalent labelling is inhibited in the presence of adenosine receptor ligands such as R-PIA.

Figure 5. Iodinated xanthine for photoaffinity crosslinking.

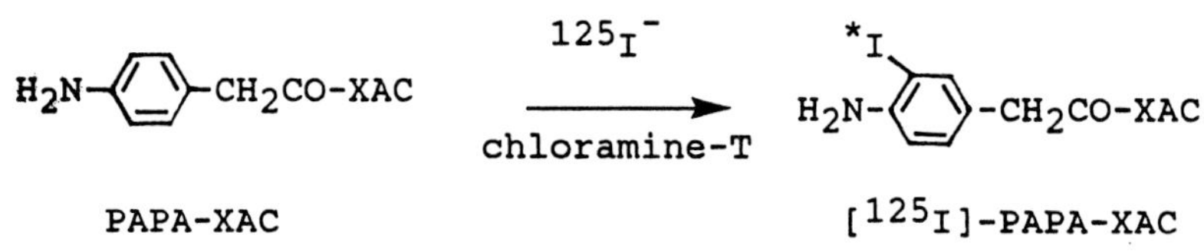

The "functionalized congener" approach has been extended to the synthesis of non-radioactive molecular probes that retain nanomolar affinity in A_1-receptor binding assays (Jacobson et al 1987a). Spectroscopic probes have been synthesized by attachment of reporter groups at distal sites of the functionalized chains of XAC and ADAC. These bifunctional ligands in the agonist series contain a spin label (eg. the nitroxyl radical TEMPO), for electron spin resonance spectroscopy, or a fluorescent dye, including fluorescein and 4-nitrobenz-2-oxa-1,3-diazole (NBD). Xanthine derivatives containing labels for ^{19}F nuclear magnetic and electron spin resonance spectroscopy have been synthesized and have high affinity at A_1-receptors.

Finally, macromolecule-bound ligands for adenosine receptors have been developed. The retention of biological activity in polymeric conjugates of adenosine receptor ligands has been used previously as evidence for the extracellular nature of the receptor. Gerlach and coworkers (Schrader et al 1977) attached adenosine nucleotides to carbonic anhydrase and were able to measure vasodilatory activity in vivo. We have utilized biotin-avidin probes to estimate binding site dimensions (Jacobson et al 1985a).

Table 4. Examples of molecular probes for A_1-receptors, binding in rat brain ([^{3}H]PIA). (Jacobson et al 1987a).

Functionalized congener/conjugate with:		Ki (nM)	probe type
ADAC	fluorescein isothiocyanate	7.1	fluorescent
ADAC	NBD chloride	4.3	fluorescent
ADAC	TEMPO isothiocyanate	4.4	spin label
XAC	TEMPO isothiocyanate	4.9	spin label
XAC	heptafluorobutyric anhydride	8.1	^{19}F NMR
XAC	diethylenetriaminepentaacetate	50	chelating

In conclusion, sites on adenosine and xanthine analogs for chain derivatization have been identified. Conjugates of these functionalized congeners display combinations of chemical, physical, and biological properties otherwise unattainable, and have been used as tools to study adenosine receptors. The possibility for drug targeting to adenosine receptors using well characterized conjugates now exists.

REFERENCES

Bruns, R.F., 1981a, Biochemical Pharmacology, 30, 325.

Bruns, R.F., Daly, J.W., & Snyder, S.H., 1980, Proceedings of the National Academy of Sciences USA, 77, 5547.

Bruns, R.F., Daly, J.W. & Snyder, S.H., 1983, Proceedings of the National Academy of Sciences USA, 80, 2077.

Evoniuk, G., Jacobson, K.A., Shamim, M.T., Daly, J.W., & Wurtman, R., 1987, Journal of Pharmacology and Experimental Therapeutics, Vol. 242, No. 3.

Fredholm, B.B., Jacobson, K.A., Jonzon, B., Kirk, K.L., Li, Y.O. & Daly, J.W., 1987, Journal of Cardiovascular Pharmacology, 9, 396.

Jacobson, K.A., Marr-Leisy, D., Rosenkranz, R.P., Verlander, M.S., Melmon, K.L. & Goodman, M., 1983, Journal of Medicinal Chemistry, 26, 492.

Jacobson, K.A., Kirk, K.L., Padgett, W. & Daly, J.W., 1985a, FEBS Letters, 184, 30.

Jacobson, K.A., Kirk, K.L., Padgett, W.L. & Daly, J.W.,

1985b, Journal of Medicinal Chemistry, 28, 1334.

Jacobson, K.A., Kirk, K.L., Padgett, W.L. & Daly, J.W., 1985, Journal of Medicinal Chemistry, 28, 1341.

Jacobson, K.A., Kirk, K.L., Daly, J.W., Jonzon, B., Li, Y-O. & Fredholm, B.B., 1985d, Acta Physiol. Scand 125:341-342.

Jacobson, K.A., Kirk, K.L., Padgett, W.L. & Daly, J.W., 1986a, Molecular Pharmacology, 29, 126.

Jacobson, K.A., Ukena, D., Kirk, K.L. & Daly, J.W., 1986b, Proceedings of the National Academy Sci. USA, 83, 4089.

Jacobson, K.A., Yamada, N., Kirk, K.L., Daly, J.W. & Olsson, R.A., 1986c, Biochemical and Biophysical Research Communications, 136, 1097.

Jacobson, K.A., Ukena, D., Kirk, K.L. & Daly, J.W., 1987a, Biochemical Pharmacology, 36, 1697.

Jacobson, K.A., Lipkowski, A.W., Moody, T.W., Padgett, W., Pijl, E., Kirk, K.L. & Daly, J.W., 1987b, Journal of Medicinal Chemistry, 30, 1529.

Jarvis, M.F., Jacobson, K.A. & Williams, M., 1987, Neuroscience Letters, in press.

Khan, M.M., Marr-Leisy, D., Verlander, M.S., Bristow, M.R., Strober, S., Goodman, M. & Melmon, K.L., 1986, Journal of Immunology, 137, 308-314.

Nye, J.S. & Snyder, S.H., 1986, Society for Neuroscience, 16th Annual Meeting, Washington, DC, Abstract 295.1.

Portoghese, P.S., 1986, Bivalent ligands in the development of selective opioid receptor antagonists. In Trends in Medicinal Chemistry: Proceedings of the Ninth International Symposium on Medicinal Chemistry, edited by E. Mutschler & E. Winterfeldt, (VCH, Weinheim, FRG) pp.327.

Rossi, N.F., Churchill, P.C., Jacobson, K.A. & Leahy, A.E., 1987, Journal of Pharmacology and Experimental Therapeutics, 240, 911.

Schrader, J., Nees, S. & Gerlach, E., 1977, Pflügers Archiv European Journal of Physiology, 369, 251.

Seale, T.W., Abla, K.A., Jacobson, K.A., Parker, K.M. & Carney, J.M., 1987, Pharmacology, Biochemistry, and Behavior, in press.

Stiles, G.L., Daly, D. & Olsson, R.A., 1985, Journal of Biological Chemistry, 260, 10806.

Stiles, G.L. & Jacobson, K.A., 1987, Molecular Pharmacology, 32, 184.

Ukena, D., Daly, J.W., Kirk, K.L. & Jacobson, K.A., 1986a, Life Sciences, 38, 797.

Ukena, D., Jacobson, K.A., Kirk, K.L. & Daly, J.W., 1986b, FEBS Letters, 199, 269.

Ukena, D., Jacobson, K.A., Padgett, W., Ayala, C., Shamim, M.T., Kirk, K.L., Olsson, R.A. & Daly, J.W., 1986c, FEBS Letters, 209, 122.

ADENOSINE ANTAGONISTS AS PHARMACOLOGICAL TOOLS

R.F. Bruns, R.E. Davis, F.W. Ninteman, B.P.H. Poschel, J.N. Wiley, and T.G. Heffner

Department of Pharmacology
Parke-Davis Pharmaceutical Research Division
Warner-Lambert Company
Ann Arbor, MI 48105 USA

INTRODUCTION

Determining the roles of adenosine *in vivo* requires the use of pharmacological tools to modulate the actions of endogenous adenosine. Of these, the receptor antagonists hold considerable promise. Although the xanthines caffeine and theophylline have long been known to block extracellular adenosine receptors (Sattin & Rall 1970), their use as tools is limited by their weak potency and their actions via other mechanisms such as phosphodiesterase inhibition and modulation of intracellular calcium. In addition, caffeine and theophylline do not distinguish between the A_1 and A_2 adenosine receptor subtypes.

Considerable effort has been invested in the search for more useful adenosine antagonists. Unfortunately, attempts to improve on the disadvantages of caffeine and theophylline have often traded away two of their outstanding advantages, namely good solubility and pharmacokinetics, resulting in loss of *in vivo* activity. The present account describes the receptor affinity, specificity, selectivity, and solubility of several adenosine antagonists, with an emphasis on potential utility for *in vitro* or *in vivo* studies. One of these agents, 8-cyclopentyltheophylline (CPT), shows potent, A_1-selective adenosine receptor blockade *in vivo* and has CNS stimulant and diuretic effects, suggesting roles for A_1 receptors in the brain and kidney.

METHODS

<u>A_1 and A_2 receptor binding assays</u>

A_1 binding affinities were determined using [^{3}H]CHA binding to rat whole brain membranes and A_2 affinities were determined using [^{3}H]NECA binding to rat striatal membranes in the presence of 50 nM N^6-cyclopentyladenosine (CPA) (Bruns *et al.* 1986).

Solubility determinations

Solubilities of poorly soluble adenosine antagonists were determined by diluting the compounds 1:100 from DMSO into Tris buffer to a final concentration above the solubility limit. The antagonist concentration in the supernatant was estimated by [^{3}H]CHA binding.

Blockade of the locomotor inhibitory effects of A_1 and A_2 agonists

Locomotor activity was measured in groups of 3 male Swiss-Webster mice (25-30 g) for 1 h in darkened cylindrical chambers containing photobeams. Drugs were dissolved or suspended by sonication in 0.9% Emulphor and were administered i.p. in a volume of 10 ml per kg body weight. CPA (2 mg/kg), 2-(phenylamino)adenosine (CV-1808, 8 mg/kg), or vehicle was administered 1 h prior to the locomotor test. CPT or vehicle was administered 30 min after the adenosine agonists, 30 min prior to testing. Five or six groups of mice were tested at each drug dose.

Blockade of the hypothermic effects of CHA

Male CF-1 Carworth mice (25-30 g) received CHA (1 mg/kg, p.o.) or vehicle (2% carboxymethylcellulose) 15 min prior to CPT (p.o.). Thirty minutes later core body temperature was measured rectally with an electronic thermistor.

Potentiation of intracranial self-stimulation

Male Long-Evans rats (450-600 g) were trained to depress a lever in order to deliver 60 Hz sine-wave alternating current (20-30 μA, 400 msec) to monopolar electrodes implanted chronically within the medial forebrain bundle and grounded to a skull screw. Under these conditions, rats display a low baseline of self-stimulation behavior. Testing was conducted by administering drugs (dissolved in DMSO vehicle) or the vehicle solution orally after collecting 1 h of baseline behavior. Drug effects were determined during each 0.5 h over 7 consecutive hours.

Natriuresis

Effects of CPT on urinary sodium excretion in hydrated rats were determined by Panlabs (Seattle, WA, USA and Taipei, Taiwan). Results are expressed as percent of vehicle (saline plus Tween 80) controls.

RESULTS AND DISCUSSION

Xanthine adenosine antagonists

Theophylline has relatively weak adenosine receptor affinity of 8.5 μM in A_1 binding and 25 μM in A_2 binding (Figure 1), but its high water solubility of 46 mM (Merck Index) results in a solubility/affinity ratio of 5,400. Theophylline inhibits phosphodiesterase in the 100 μM to 1 mM concentration range, so its specificity for adenosine blockade relative to phosphodiesterase inhibition is only about 10-fold. It is very

Figure 1. Structures, A_1 and A_2 binding affinities, and solubilities of xanthine adenosine antagonists.

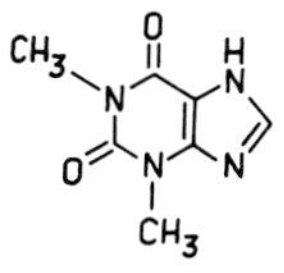

theophylline

A_1	8,500 nM
A_2	25,000 nM
A_2/A_1	3.0
solubility	46 mM
solubility/A_1	5,400

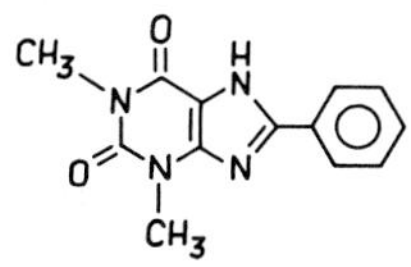

8-phenyltheophylline

A_1	86 nM
A_2	850 nM
A_2/A_1	9.9
solubility	7.8 μM
solubility/A_1	90

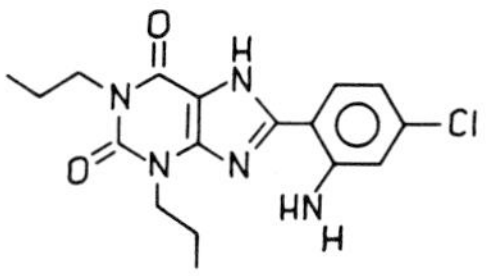

PACPX

A_1	2.5 nM
A_2	92 nM
A_2/A_1	37
solubility	210 nM
solubility/A_1	84

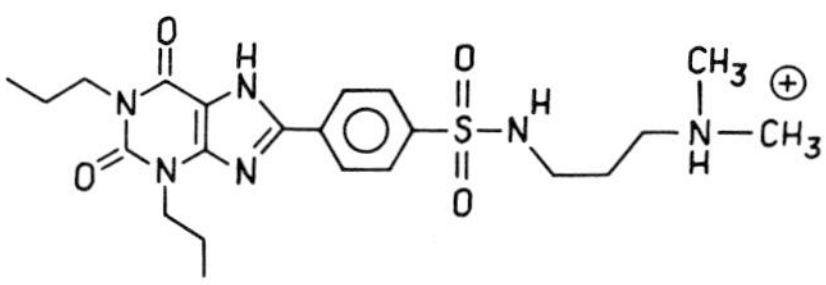

PD 113,297

A_1	5.6 nM
A_2	70 nM
A_2/A_1	12.5
solubility	800 μM
solubility/A_1	140,000

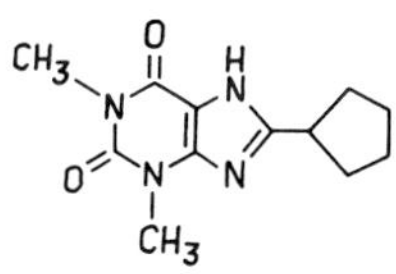

CPT

A_1	11 nM
A_2	1,400 nM
A_2/A_1	130
solubility	390 μM
solubility/A_1	35,000

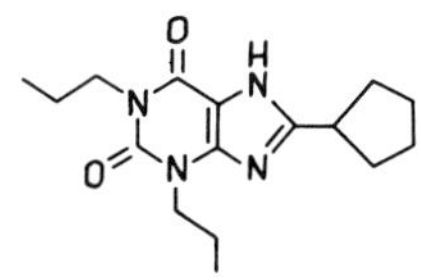

CPX

A_1	0.46 nM
A_2	340 nM
A_2/A_1	740
solubility	9.9 μM
solubility/A_1	21,000

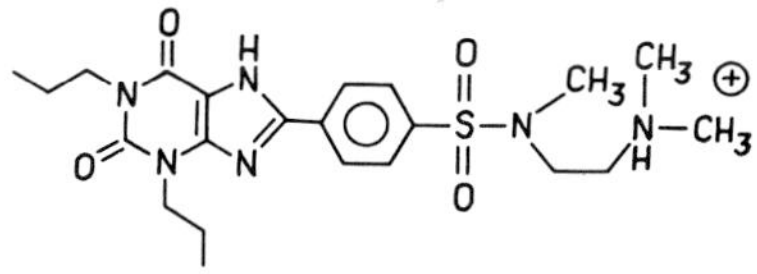

PD 115,199

A_1	14 nM
A_2	16 nM
A_2/A_1	1.1
solubility	35 μM
solubility/A_1	2,500

slightly A_1-selective (3-fold) in the two binding assays that we use as benchmarks.

The first potent adenosine antagonist, 8-phenyltheophylline (8-PT) (Bruns 1981), is about 100-fold more potent than theophylline in the rat brain A_1 binding assay (Figure 1), but unfortunately it is also about 6000-fold less soluble (solubility 7.8 μM). Despite the unfavorable solubility/affinity ratio of 90 for this compound, it has proven useful in many *in vitro* studies and in some *in vivo* studies as well (Lautt 1987; Collis *et al.* 1986). 8-PT does not inhibit phosphodiesterase up to its solubility limit (Smellie *et al.* 1979).

8-(2-Amino-4-chlorophenyl)-1,3-dipropylxanthine (PACPX) was the end result of an attempt to maximize the affinity of the 8-PT series, without regard to solubility (Bruns *et al.* 1983). Although PACPX is considerably more potent than 8-PT in A_1 binding (Figure 1), it has about the same ratio of solubility to affinity as 8-PT, and has not generally proven very useful as a pharmacological tool. PACPX is moderately A_1-selective (37-fold) in the rat.

One potential strategy for improving solubility is to add a charged side-chain to the parent 8-phenylxanthine structure. Among the compounds resulting from this strategy are the xanthine congeners XCC and XAC (Jacobson *et al.* 1985) and a series of amino-substituted xanthine-sulfonamides exemplified by PD 113,297 (Figure 1) (Hamilton *et al.* 1985). A fortuitous result with the amine-containing analogs such as XAC and PD 113,297 was an increase in adenosine receptor affinity, probably due to an ionic interaction between the positively charged amine and a distal negative charge on the receptor. PD 113,297 has a very high solubility/affinity ratio of 140,000. It is a potent adenosine antagonist in *in vitro* preparations such as the rat isolated heart (S. Haleen and R. Steffen, personal communication) and the guinea pig ileum (L. Gustafsson, personal communication), but its *in vivo* activity has so far proved disappointing. PD 113,297 is only about 8-fold more potent than theophylline in blocking A_1 and A_2 challenges when given intravenously (T. Mertz, personal communication), and is inactive by the i.p. route in the motor activity model for A_1 and A_2 blockade (see Methods). The reason for the poor activity of PD 113,297 *in vivo* is not clear; one possibility is that the addition of a charged moiety at one end of a very hydrophobic structure results in an amphiphilic molecule that could be sequestered in membranes or other hydrophobic sites.

8-Cyclopentyltheophylline (CPT) is the first adenosine antagonist with a useful degree of A_1-selectivity (130-fold in the rat) (Bruns *et al.* 1986). The high A_1 affinity (K_i 10 nM) of this compound, coupled with its moderately good water solubility, results in a high solubility/affinity ratio of 35,000 (Figure 1). CPT also has an octanol:water partition coefficient (log P) of 2.0, which is close to ideal for *in vivo* activity. As shown later in this paper, CPT is a potent, orally active, A_1-selective adenosine antagonist *in vivo*. CPT is a weak phosphodiesterase inhibitor, with an IC50 of about 200 μM (Goodsell *et al.*, 1970).

The next logical step after CPT is the propyl homolog 8-cyclopentyl-1,3-dipropylxanthine (CPX; also referred to as DPCPX or PD 116,948). Dipropyl substitution gives a boost of 25-fold in A_1 affinity (to 0.46 nM) and 6-fold in A_1-selectivity (to 740-fold), resulting in the most potent and selective antagonist available to date. The solubility of CPX is 9.9 μM. Although the relatively low solubility and high log P (3.6) of CPX initially led us to take a pessimistic view of its prospects for *in vivo* activity (Bruns *et al.* 1987a), later results indicate that CPX is considerably more potent and A_1-selective than CPT in the motor activity model (A_1 ED50 0.2 mg/kg i.p.; A_2 >30 mg/kg). CPX should be a very useful pharmacological tool for *in vitro* studies; for instance, in the rat isolated heart it shows about 1000-fold greater potency against the heart-rate lowering (A_1) effects of R-PIA than against R-PIA's vasodilating (A_2) effects (Haleen *et al.* 1987). [^{3}H]CPX is also an excellent antagonist ligand for A_1 receptors, showing only about 1% nonspecific binding in rat brain membranes (Bruns *et al.* 1987a).

Progress towards A_2 antagonist xanthines has been much slower than towards A_1 blockers. A first step in the direction of selective A_2 antagonists is PD 115,199 (Figure 1), a member of the same series as PD 113,297. Although the other members of this series were weaker in A_2 binding than A_1, PD 115,199 had equally high affinity in both assays (A_1 K_i 14 nM, A_2 K_i 16 nM) (Bruns *et al.* 1987b). PD 115,199 shows about 10-fold A_2-selectivity in dog and human A_1/A_2 binding. As yet, little information is available about the use of PD 115,199 as a pharmacological tool *in vitro* or *in vivo*. [^{3}H]PD 115,199 has been used as a radioligand; in the presence of 20 nM unlabeled CPX (to eliminate A_1 binding) [^{3}H]PD 115,199 labels a high-affinity A_2 receptor in rat striatal membranes (Bruns *et al.* 1987b). About 80% of [^{3}H]PD 115,199 binding is specific.

Non-xanthine adenosine antagonists

Non-xanthine adenosine antagonists are increasingly being reported. At present these agents represent chemical leads rather than pharmacological tools (an exception may be the A_2-selective antagonist CGS 15943 reported by Williams *et al.* (1987)), but over the long run they may provide important advantages over the xanthines. Several of the leads that we have found are described below.

N^6-Butyl-8-phenyladenine (Figure 2) was recently reported to have high affinity in adenosine receptor binding (Bruns & Coughenour 1987). Since the structure and affinity of this compound bear some resemblance to 8-PT, it is tempting to speculate that the binding modes for the two compounds will be similar.

Davies *et al.* (1983) reported that several 4-thiopyrazolo[3,4-*d*]pyrimidines had significant adenosine antagonist activity. We

Figure 2. Structures, A_1 and A_2 binding affinities, and solubilities of non-xanthine adenosine antagonists.

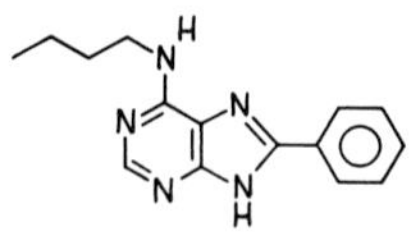

N^6-butyl-8-phenyladenine

A_1	173 nM
A_2	661 nM
A_2/A_1	3.8
solubility	5.4 µM
solubility/A_1	31

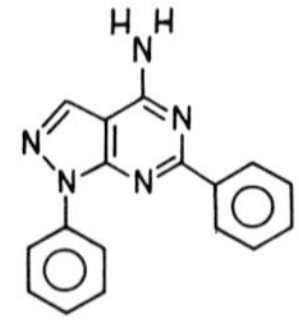

APPP

A_1	23 nM
A_2	35 nM
A_2/A_1	1.5
solubility	13 µM
solubility/A_1	570

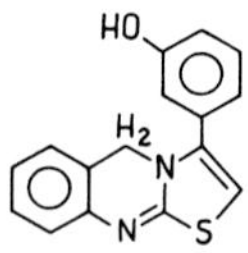

HTQZ

A_1	3100 nM
A_2	120 nM
A_2/A_1	0.04
solubility	14.5 µM
solubility/A_2	121

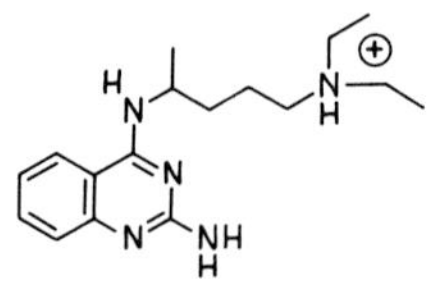

ADQZ

A_1	610 nM
A_2	330 nM
A_2/A_1	0.55
solubility	280 µM
solubility/A_2	850

independently found A_1 and A_2 affinity in a series of 4-aminopyrazolo[3,4-*d*]pyrimidines that had been sent to Parke-Davis for screening in 1960 by Dr. E.C. Taylor (Princeton University) (Bruns & Hamilton, 1987). The most potent of this series, 4-amino-1,6-diphenylpyrazolo[3,4-*d*]pyrimidine (APPP) (Figure 2), had K_i values of 23 nM in A_1 binding and 35 nM in A_2 binding. However, solubility was relatively low at 13 µM. Because of its poor solubility, this compound has not been extensively characterized in functional antagonist assays.

3-(3-Hydroxyphenyl)-5H-thiazolo[2,3-*b*]quinazoline (HTQZ) (Figure 2) was originally identified from screening of selected compounds from the Parke-Davis compound bank in [^{3}H]CHA binding (Bruns & Coughenour, 1987). Although the A_1 affinity of this compound was not impressive (K_i 3.1 µM), it was later found to have high affinity (K_i 120 nM) in the A_2 binding assay. The 25-fold A_2-selectivity shown by HTQZ is the greatest yet reported for any adenosine antagonist or agonist in the A_1/A_2 binding assays. Despite this, the usefulness of HTQZ as a pharmacological tool is limited by several factors. Solubility of HTQZ is poor (14 μM), and the A_2-selectivity of HTQZ is much lower in dog, human, and bovine brain than in rat brain.

The high A_2 affinity of the fused quinazoline HTQZ inspired us to screen simple quinazolines for A_1 and A_2 affinity (R.F. Bruns & S.A. Johnson, manuscript in preparation). A potentially interesting chemical lead that we found in this manner was 2-amino-4-{[(4-(diethylamino)-1-methylbutyl]amino}quinazoline (ADQZ) (Figure 2). In addition to its moderately potent A_2 affinity of 330 nM and its 2-fold A_2-selectivity, ADQZ is of interest because it contains an aminoalkyl side-chain instead of the aryl or cycloalkyl group found in all of the other adenosine antagonists with nanomolar affinity. ADQZ has moderate solubility (280 μM).

CNS adenosine antagonism *in vivo*

In view of the poor *in vivo* activity of adenosine antagonists such as 8-PT and PACPX, we sought to establish a rational screening strategy capable of identifying the degree of *in vivo* antagonism exhibited by different compounds. Four different screens were utilized to pinpoint which of the requirements for *in vivo* activity might be lacking in a given compound. The rat Langendorff preparation was used to measure functional antagonism *in vitro* (Haleen *et al.* 1987), followed by an intravenous cardiovascular screen designed to detect potential problems in systemic distribution of antagonists (T. Mertz, manuscript in preparation). Finally, the two whole animal screens described below were used to determine antagonist activity by the i.p. and oral routes.

Antagonism by CPT of the A_1 and A_2 inhibition of motor activity. The ability of adenosine antagonists to reverse the motor inhibitory effects of the A_1 agonist CPA (2 mg/kg i.p.) and the A_2 agonist CV-1808 (8 mg/kg i.p.) was investigated (Figure 3). Antagonists were given i.p. 30 minutes after the agonists. CPT blocked the motor activity decrease produced by the A_1 agonist CPA (ED50 10 mg/kg) but did not affect the response to the A_2 agonist CV-1808. CPX was even more potent and selective (ED50 0.2 mg/kg versus CPA, >30 mg/kg versus CV-1808). However, PD 113,297 was inactive at 30 mg/kg in this test. These results indicate that CPT and CPX are absorbed by the i.p. route and are able to cross the blood-brain barrier. In addition, the selectivity for the A_1 receptor shown by the two antagonists in binding studies carries over to the whole animal.

Antagonism by CPT of A_1-elicited hypothermia. Adenosine agonists cause marked hypothermia in small animals such as mice (Mehta & Kulkarni, 1983). We used this paradigm to measure A_1 blockade by the oral route. Adenosine antagonists were given orally 15 minutes after CHA (1 mg/kg p.o.) and core temperature was measured

Figure 3. Reversal of the motor inhibitory effects of CPA (an A_1 agonist) and CV-1808 (an A_2 agonist) by CPT.

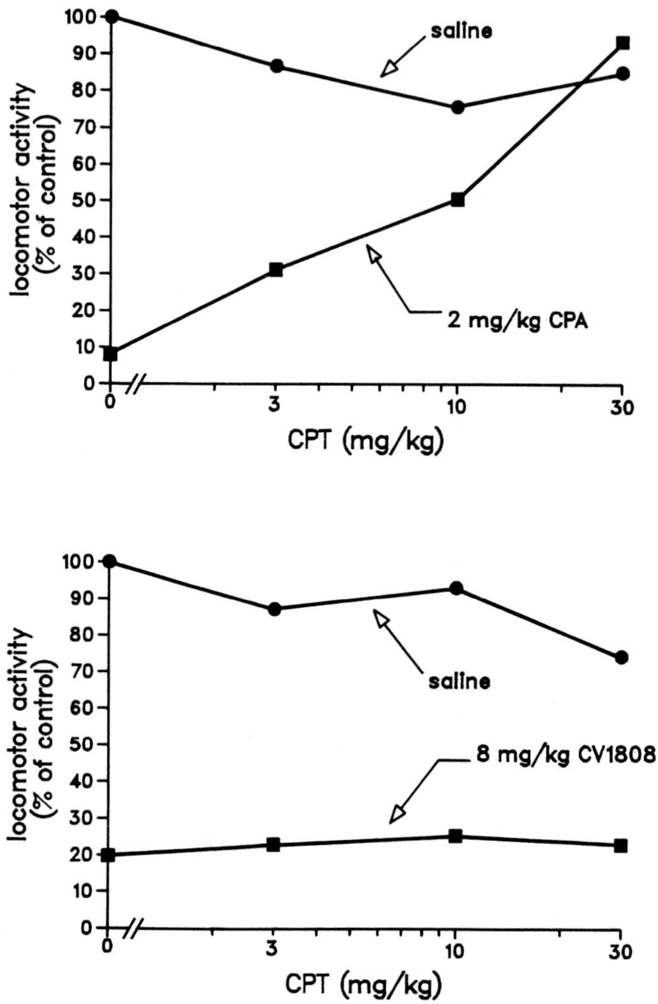

30 minutes later. CHA caused a 4°C drop in body temperature, which was inhibited more than 50% by 1 mg/kg of CPT (Figure 4), indicating that CPT is a potent A_1 antagonist by the oral route. In contrast, PD 113,297 was inactive in this test at 30 mg/kg.

Figure 4. Inhibition by CPT of the hypothermia elicited by the A_1 agonist CHA.

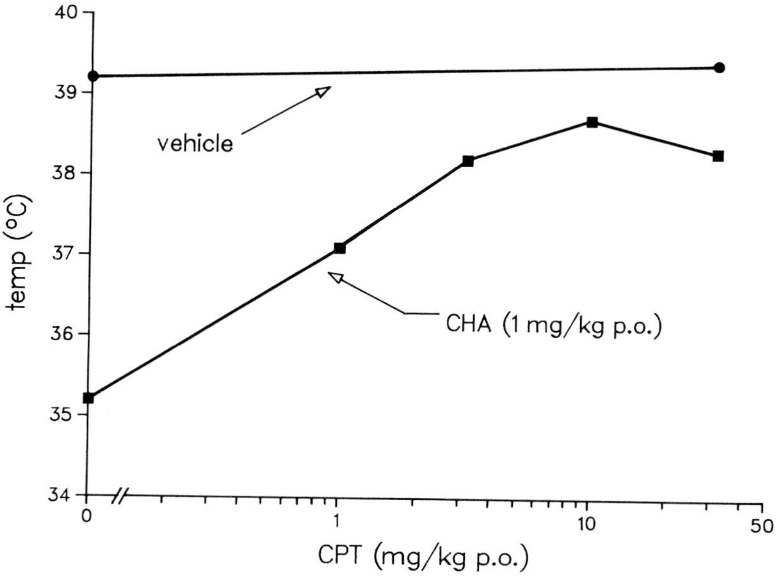

CNS and renal actions of CPT *in vivo*

If adenosine has a significant role in the normal animal, then specific adenosine antagonists should have effects by themselves. We have used CPT as a tool to begin studying this question.

Effects of CPT on self-stimulation. We have already seen (Figure 3) that CPT when given alone has little effect on motor activity. In contrast, theophylline (10 mg/kg) causes an 80-100% increase in motor activity in the same test. These data indicate that a specific A_1-selective adenosine antagonist will not share the overt stimulant profile of theophylline, suggesting that at least part of the stimulant activity of caffeine and theophylline may be due to A_2 blockade or to non-adenosine effects. However, CPT does have more subtle CNS stimulant effects. For instance, we have noted an exaggerated startle reflex in animals given 50 mg/kg p.o. of CPT. Another relevant test is the self-stimulation paradigm, which measures the ability of drugs to enhance operant responding to obtain electrical stimulation of the medial forebrain bundle. This test is considered a predictor of CNS stimulant or antidepressant activity. CPT caused a 6-fold increase in self-stimulation at 12.5 mg/kg p.o., the lowest dose tested (Figure 5). Taken

Figure 5. Potentiation of intracranial self-stimulation by CPT.

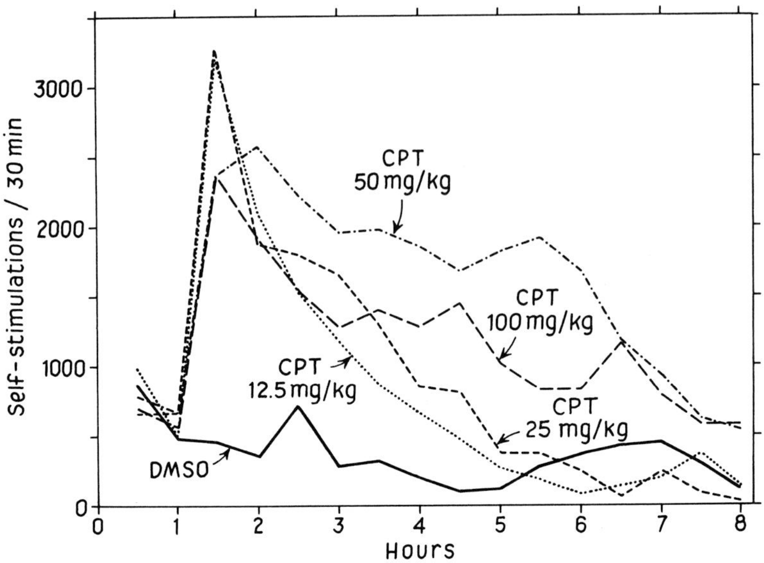

Figure 6. Stimulation of urinary sodium excretion by CPT.

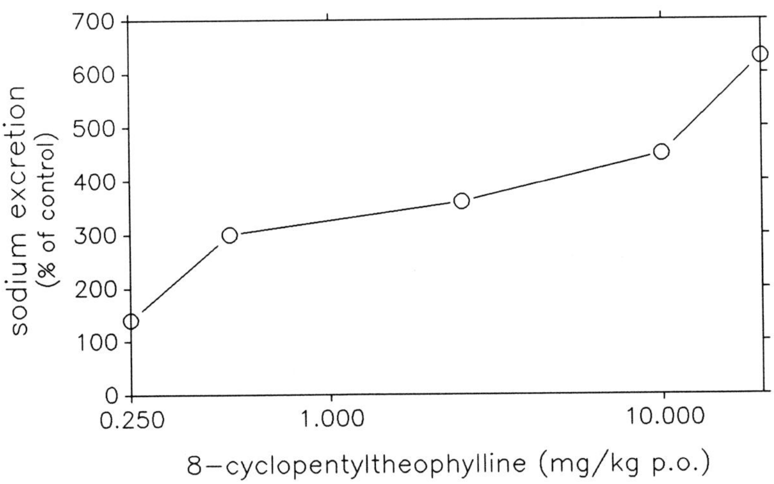

together, these results suggest that A_1 antagonism is responsible for a subset of the CNS effects of caffeine and theophylline.

Effects of CPT on renal sodium excretion. The role of adenosine in tubuloglomerular feedback and other renal functions is well-known (Osswald 1984). Caffeine and theophylline are diuretics, and Collis *et al.* (1986) have shown that 8-PT has diuretic activity. As part of the characterization of CPT, we submitted this compound to Panlabs for evaluation in an assortment of screens. Potent natriuretic activity was seen, with significant enhancement of sodium excretion at 0.5 mg/kg p.o. and a maximum increase of 6-fold (Figure 6). We have also seen diuretic activity in several other balanced or A_1-selective adenosine antagonists, including several non-xanthines, implying that this effect is probably due to adenosine antagonism rather than to some extraneous effect of the xanthines unrelated to adenosine.

SUMMARY

In these studies, we have identified adenosine antagonists with the necessary potency, selectivity, and solubility to be of use as pharmacological tools. CPT was shown to be a potent, orally active, A_1-selective antagonist *in vivo*. The activity of CPT by itself in the CNS and kidney suggests ongoing roles for A_1 receptors in these tissues, and suggests potential utility of A_1-selective adenosine antagonists as antidepressants or diuretics.

REFERENCES

Bruns, R.F., 1981, Biochemical Pharmacology, 30, 325.

Bruns, R.F. & Coughenour, L.L., 1987, The Pharmacologist, 29, 146.

Bruns, R.F. & Hamilton, H.W., 1987, Federation Proceedings, 46, 393.

Bruns, R.F., Daly, J.W. & Snyder, S.H., 1983, Proceedings of the National Academy of Sciences USA, 80, 2077.

Bruns, R.F., Lu, G.H. & Pugsley, T.A., 1986, Molecular Pharmacology, 29, 331.

Bruns, R.F., Fergus, J.H., Badger, E.W., Bristol, J.A., Santay, L.A., Hartman, J.D., Hays, S.J. & Huang, C.C., 1987a, Naunyn-Schmiedeberg's Archives of Pharmacology, 335, 59.

Bruns, R.F., Fergus, J.H., Badger, E.W., Bristol, J.A., Santay, L.A. & Hays, S.J., 1987b, Naunyn-Schmiedeberg's Archives of Pharmacology, 335, 64.

Collis, M.G., Baxter, G.S. & Keddie, J.R., 1986, Journal of Pharmacy and Pharmacology, 38, 850.

Davies, L.P., Brown, D.J., Chen Chow, S., & Johnston, G.A.R., 1983, Neuroscience Letters, 41, 189.

Goodsell, E.B., Stein, H.H. & Wenzke, K.J., Journal of Medicinal Chemistry, 14, 1202.

Haleen, S.J., Steffen, R.P. & Hamilton, H.W., 1987, Life Sciences, 40, 555.

THE PHARMACOLOGICAL CLASSIFICATION AND STRUCTURE OF PERIPHERAL ADENOSINE RECEPTORS

DAVID M PATON
Department of Pharmacology and Clinical Pharmacology,
University of Auckland,
Auckland, New Zealand

INTRODUCTION

Adenosine and adenosine 5'-triphosphate (ATP) have numerous effects on peripheral tissues. These have been the subject of a number of recent reviews (see, for example, Stefanovich et al. 1985, Paton 1985). The present article reviews the actions of adenosine in causing inhibition of transmission in peripheral adrenergic and cholinergic nerves. Particular emphasis has been given to the question of the classification of the receptor sites responsible for this action and to the structural requirements for activity at adenosine receptors at these sites.

Within the central nervous system, adenosine also reduces the release of many transmitters including that of acetylcholine, 5-hydroxytryptamine, dopamine, GABA and glutamate, and produces presynaptic inhibition of neurotransmitter release at the peripheral neuromuscular junction (Dunwiddie, 1985).

METHODS AND PREPARATIONS USED

In all the studies conducted in this laboratory, we have utilised the mid-portion of the rat vas deferens (for studies of adrenergic transmission) and ileal segments from guinea-pig (for studies of cholinergic transmission). These tissues are suspended longitudinally and contractility recorded isometrically. Twitch responses of the preparations are produced by continuous field stimulation at 0.2 Hz (1ms duration and supramaximal voltage). Under these conditions, addition of adenosine and of a large number of adenosine analogues or of the 5'-adenine nucleotides causes rapid inhibition of the twitch responses, this effect being antagonised by 30μM theophylline. On washing the tissue, twitch responses to field stimulation recover, the time required for recovery varying with the different analogues studied. The % inhibition of responses is determined and the concentration producing 50% inhibition of the maximal inhibition (EC_{50}) determined. In these tissues, inhibition of twitch responses occurs through presynaptic inhibition of transmitter release (Paton 1981).

THE CLASSIFICATION OF ADENOSINE RECEPTORS

Adenosine receptors have been classified into two subtypes based on the effects of adenosine and adenosine analogues on the activity of the enzyme, adenylate cyclase. Activation of one such sub-type results in inhibition of the activity of the enzyme and these adenosine receptors are termed A_1 (or R_i) receptors.

Activation of another sub-type results, in contrast, in activation of adenylate cyclase activity and such receptors are termed A_2 or (R_a) receptors (Paton 1985).

The classification of adenosine receptors in many tissues has proved more difficult and problematical because physiological or biochemical responses have generally had to be measured, while changes in the activity of adenylate cyclase have not been or could not be monitored simultaneously in these studies. In addition, there has been until recently a lack of relatively selective adenosine antagonists and this has also hindered attempts at such classification. Consequently, reliance has had to be placed very largely on the relative order of potencies of agonists (Paton 1985). The problems associated with this approach and the underlying assumptions made have been carefully reviewed by Collis (1985).

Very often, the classification of adenosine receptors in peripheral tissues in which physiological responses have been monitored, has been based on the rank order of potencies of R-PIA, S-PIA and NECA. However, in several such studies, the results obtained using NECA have yielded rather equivocal results (see for example Collis 1985). An early example of this was that reported in our own studies of guinea-pig ileum and rat vas deferens (Paton 1981).

In an attempt to clarify and hopefully to simplify the classification of adenosine receptors in tissues in which physiological responses are monitored, Ukena et al. (1987) have examined the relative potencies of 145 N^6-substituted adenosine analogues as agonists in a variety of tissues. As a result of this and earlier studies by the same workers (Daly et al. 1986), these investigators proposed that a set of 31 analogues may be useful for such classification purposes and further that a sub-set of 7 analogues, namely

NECA, N^6-cyclohexylNECA, N^6-CHA, R-PIA, S-PIA, 2-chloroadenosine and 2-(phenylamino)adenosine, might be particularly valuable.

In the present instance, an attempt has been made to define an even more restricted number of adenosine analogues for the classification of adenosine receptors. The following considerations have been involved in this endeavour: (a) the likely availability of such analogues, (b) the selection of analogues with substitutions at either the N^6, or C^2 or 5' positions, and (c) a consideration of known, reported differences in relative potencies at putative A_1 and A_2 receptors as defined in previously published radioligand binding studies. For this latter consideration, the data presented by Bruns et al. (1986) on the inhibition of [^{3}H]CHA binding in rat brain and for the inhibition of binding of [^{3}H]NECA in rat striatum has been particularly valuable.

The potential value of several sets of analogs was studied by determining correlation coefficients for their activity at adenosine receptors utilising data reported for 9 different tissues as well as some unpublished data from this laboratory for guinea-pig ileum and rat vas deferens. Table I shows the results obtained using the four analogues N^6-cyclopentyladenosine, R-PIA, NECA, and 2-(phenylamino)-adenosine. Five of the tissues used have been considered in previous studies to have adenosine receptors of the A_1 sub-type. These are rat brain (inhibition of [^{3}H]CHA binding), rat fat cells (inhibition of adenylate cyclase activity), rat hippocampus (inhibition of firing of neurones), guinea-pig ileum (inhibition of transmitter release) and rat vas deferens (inhibition of transmitter release). The other four tissues have been considered to have adenosine receptors belonging to the A_2 sub-type; these tissues are rat striatum, (inhibition of [^{3}H]NECA binding), rat PC12 cells (activation of adenylate cyclase activity), human platelets (activation of adenylate cyclase activity) and dog coronary artery (vasodilation).

Correlation coefficients ranged from 0.66 - 1.00 within the first group of tissues with putative A_1 receptors and from 0.80 - 0.98 within the second group of tissues with putative A_2 receptors, while correlation coefficients <u>between</u> the two groups ranged from -0.01 to -0.78 (Table 1). The only tissue which had correlation coefficients that differed from these was the rat vas deferens (putative A_1 receptors) where correlation coefficients to tissues of the A_2 sub-type ranged from -0.07 to 0.50.

It can thus be seen that the use of these four analogues has allowed a very clear separation of two groupings or families or sub-types of receptors in these two groupings of tissues, thus supporting previous conclusions that these tissues can indeed be classified as either A_1 or A_2 in type.

As suggested by earlier work (Paton 1981), it seems possible that the adenosine receptors present in rat vas deferens may well be predominantly A_1 in type, but that in addition a proportion of A_2 receptors may also be present in this tissue. Clearly the use of agonists alone will never allow one to exclude the possibility that a mixture of adenosine receptors is present in any particular tissue. This possibility will need to be further explored by the use of more selective antagonists as these become available and by the use of potent radioligands for the adenosine receptors. The value of this approach was recently shown by the demonstration of A_1 adenosine receptors on rat ventricular myocytes using the A_1 selective radioligand R-N^6-[^{125}I]-p-hydroxyphenylisopropyladenosine (Martens et al. 1987), thus confirming the conclusions of earlier physiological studies (Kurahashi and Paton 1982).

AN APPROACH TO THE STRUCTURE OF PERIPHERAL PRESYNAPTIC ADENOSINE RECEPTORS

The molecular nature of the peripheral adenosine receptor in adrenergic and cholinergic nerves will ultimately be determined using the type of biochemical approaches outlined elsewhere in this volume by Stiles. However, in the interim, indirect approaches are useful in order to obtain some idea of at least portions of the adenosine receptors in such tissues. For this purpose a detailed study has been made of the structural requirements for activity at the N^6 region of the adenosine receptors present in both guinea-pig ileum and rat vas deferens, utilising the physiological responses of these tissues to a large series of adenosine analogues (Paton 1981, Paton and Webster 1984, Paton and Olsson 1985, Paton et al. 1986, 1987).

The starting point for this series of studies has been the observation of the marked potency of R-PIA and the marked stereoselectivity of R-PIA relative to S-PIA in these tissues (Paton 1981). This has led to the testing of a model of the N^6 region of the adenosine receptor region in which this region is considered to be complementary to that of the N^6 substituent of R-PIA (Daly et al. 1986, Paton et al. 1986).

In the case of R-PIA and S-PIA, the carbon propyl C2 is the chiral centre and the positions of propyl C1 carbon and

Table 1 Correlation coefficients for relative potencies of four adenosine agonists in selected tissues

	Correlation Coefficients (r)								
	Brain (a)	Fat cells (b)	Hippocampus (c)	Ileum (d)	Vas deferens (e)	Striatum (f)	PC12 cells (g)	Platelets (h)	Coronary artery (i)
Brain		1.00	0.99	0.95	0.71	-0.22	-0.42	-0.56	-0.75
Fat cells	1.00		0.99	0.92	0.66	-0.26	-0.46	-0.60	-0.78
Hippocampus	0.99	0.99		0.92	0.74	-0.12	-0.32	-0.49	-0.69
Ileum	0.95	0.92	0.92		0.86	-0.01	-0.22	-0.34	-0.55
Vas deferens	0.71	0.66	0.74	0.86		0.50	0.31	0.18	-0.07
Striatum	-0.22	-0.26	-0.12	-0.01	0.50		0.98	0.92	0.80
PC12 cells	-0.42	-0.46	-0.32	-0.22	0.31	0.98		0.98	0.91
Platelets	-0.56	-0.60	-0.49	-0.34	0.18	0.92	0.98		0.97
Coronary artery	-0.75	-0.78	-0.69	-0.55	-0.07	0.80	0.91	0.97	

Correlation coefficients were determined for the relative potencies of N^6-cyclopentyladenosine, R-PIA, 2-(phenylamino) adenosine and NECA reported for: (a) inhibition of $[^3H]N^6$-cyclohexyladenosine binding in rat brain (Bruns et al. 1986); (b) inhibition of adenylate cyclase in rat fat cell membranes (Ukena et al. 1987); (c) inhibition of firing in rat hippocampus (Dunwiddie et al. 1986); inhibition of twitch responses of (d) guinea-pig ileum and (e) rat vas deferens (Paton et al. 1986); (f) inhibition of $[^3H]$NECA binding in rat striatal membranes (Bruns et al. 1986); stimulation of adenylate cyclase in (g) rat PC12 cells and (h) human platelet membranes (Ukena et al. 1987); (i) vasodilation in dog coronary artery (Daly et al. 1986).

C3 define the absolute configurations. The sub-regions of the receptor that interact with C1, C2 and C3 carbons are designated as S-1, S-2 and S-3 sub-regions respectively. The receptor sub-region for interaction with the proton on the chiral carbon of R-PIA is designated S-4. In addition to this, an aryl (phenyl) sub-region may also be adjacent to the S-1 sub-region.

A study of straight chain N^6 alkyl substituents showed that either the S-1 or S-3 sub-regions are large enough to accommodate an ethyl group. In addition, all the secondary alkyl compounds were more potent than the corresponding primary alkyl analogues, indicating that simultaneous interactions with both the S-1 and S-3 sub-region enhanced potency. Ethyl groups were optimal for the interactions at both the S-1 and S-3 sub-regions.

Replacing a methyl group with a phenyl group either had little effect or reduced potency. At these receptor sites therefore, a sub-region for an aryl substituent (e.g. phenyl) may not be a prominent feature.

Stereoselectivity is an important feature at this site as demonstrated by the considerably greater potency of R-PIA compared to S-PIA. This also holds true for the R and S diastereomers of N^6-(1-phenyl-2-butyl)adenosine. This marked stereoselectivity could result from a positive interaction between the propyl C3 methyl or ethyl groups and the S-3 sub-region, or a negative interaction of the C3 methyl groups with the S-4 sub-region, or a combination of the two. A study with diastereomeric analogues capable of interacting with either the S-3 or S-4 sub-regions showed that the R-diastereomers were more potent and the S-diastereomers were less potent than their primary alkyl analogues. Tertiary alkyl analogues of these compounds had generally lower activity. The marked stereoselectivity at the C2 chiral centre can therefore be accounted for by a combination of both a positive interaction of the propyl C3 of R-diastereomers with the S-3 sub-region as well as by a negative interaction of propyl C3 of S-diastereomers with the S-4 sub-region.

An unpublished study of the series of tertiary alkyl congeners relative to their secondary alkyl analogues has provided further information about the S-4 sub-region. In this series it was found that if the tertiary alkyl substituent is small (e.g. methyl), the number of methyl plus methylene groups interacting with the receptor appears to change relatively little and potency decreases little if at all. However, if the tertiary alkyl substituent is larger (e.g. ethyl) the interaction at the putative S-4 sub-region may position the rest of the N^6 substituent so that interaction with the receptor is no longer possible or optimal, causing a consequent large decrease in potency of the agonist. This may result from steric hindrance.

The ability of a series of N^6-modified N-alkyl-5'-uronamides to cause inhibition of transmitter release in guinea-pig ileum has also been examined. It was found that these agents were indeed agonists, the majority being full agonists. N-ethyl 5'-uronamide substitution resulted in up to 4-fold reduction in activity of N^6-substituted adenosine analogues, while stereoselectivity of the N^6-substituted analogues continued to be present. 5-Uronamide substitutions to N^6-(3-pentyl)-adenosine resulted in a marked loss of activity when there were large alkyl groups at the amide or with amides of secondary amines. It was concluded that adenosine analogues apparently interact with both the N^6 and C-5' sub-regions of the adenosine receptor in this tissue, the interaction however being less than additive in this instance.

These features of the adenosine receptor at the N^6 region of the adenosine receptor site are generally similar to those described for the A_1 adenosine receptor studied in rat hippocampus (inhibiting of neuronal firing, Dunwiddie et al. 1986), fat cells (inhibition of adenylate cyclase activity, Ukena et al. 1987) and brain (inhibition of [^{3}H]CHA binding, Ukena et al. 1987). It thus seems possible that this portion of the A_1 adenosine receptor may be structurally rather similar in different tissues. However, proof of this will await more sophisticated biochemical and molecular studies of the isolated adenosine receptor in these tissues.

ACKNOWLEDGEMENTS

Studies in the author's laboratory were supported by the Medical Research Council of New Zealand.

REFERENCES

Bruns, R.F., Lu, G.H. and Pugsley, T.A. 1986, Molecular Pharmacology, 29, 331-346.

Collis, M.G., 1985, in Purines: Pharmacology and Physiological Roles, edited by T.W. Stone (Macmillan, London), p. 75-84.

Daly, J.W., Padgett, W., Thompson, R.D., Kusachi, S., Bugni, W.J. and Olsson, R.A., 1986, Biochemical Pharmacology, 35, 2467-2481.

Dunwiddie, T.W., 1985, International Review of Neurobiology, 27, 63-139.

Dunwiddie, T.W., Worth, T.S and Olsson, R.A., 1986, Naunyn-Schmiedeberg's Archives of Pharmacology, 334, 77-85.

Kurahashi, K. and Paton, D.M., 1982, British Journal of Pharmacology, 74, 829P - 830P.

Martens, D., Lohse, M.J., Rauch, B. and Schwabe, U., 1987, Naunyn-Schmiedeberg's Archives of Pharmacology, 336, 342-348.

Paton, D.M., 1981, Journal of Autonomic Pharmacology, 1, 278-290.

Paton, D.M., 1985, editor Methods in Pharmacology, Vol. 6 (Plenum Press, New York).

Paton, D.M., Lockwood, P.A., Martlew, D.L. and Olsson, R.A., 1987, Naunyn-Schmiedeberg's Archives of Pharmacology, 335, 301-304.

Paton, D.M. and Olsson, R.A., 1985, in Purines: Pharmacology and Physiological Roles, edited by T.W. Stone (Macmillan, London). p. 85-94.

Paton, D.M., Olsson, R.A. and Thompson, R.D., 1986, Naunyn-Schmiedeberg's Archives of Pharmacology, 333, 313-322.

Paton, D.M. and Webster, D.R., 1984, in Extraneuronal Events in Autonomic Pharmacology, edited by W.W. Fleming, S.Z. Langer, K.H. Graefe and N. Weiner (Raven Press, New York), p. 193-204.

Stefanovich, V., Rudolphi, K., and Schubert, P., 1985, editors Adenosine: Receptor and Modulation of Cell Function (IRL Press, Oxford).

Ukena, D., Olsson, R.A. and Daley, J.W., 1987, Canadian Journal of Physiology and Pharmacology, 65, 365-376.

THE STRUCTURE AND REGULATION OF THE A_1 ADENOSINE RECEPTOR

Gary L. Stiles*

Departments of Medicine and Biochemistry
Duke University Medical Center
Durham, North Carolina

I. INTRODUCTION

Adenosine acting via A_1 adenosine receptors(A_1AR) is pable of inducing a wide range of physiological effects cluding suppression of cardiac function, induction of dation and inhibition of lipolysis. Although many of ese effects are believed to be mediated via inhibition of e enzyme adenylate cyclase, A_1 adenosine receptors are so coupled to effector systems other than adenylate clase, such as the K^+ channel (Stiles 1986). The A_1 enosine receptor can be classified pharmacologically by a ique potency series of adenosine analogs such that R-PIA -phenylisopropyladenosine) is more potent than NECA '-N-ethylcarboxamide adenosine) which is more potent than PIA (Stiles 1986).

Our understanding of the structure and function of enosine receptors has evolved through three technological ases. The first was the use of physiological studies ich demonstrated that there were two types of adenosine ceptors termed A_1 (R_i) and A_2 (R_a). A_1 receptors inhibit ile A_2 receptors stimulate adenylate cyclase. These ceptors are nonhomogeneously distributed throughout the sues of the body and demonstrate distinct pharmacologic ecificities and differential affinities for various ectors. The second phase was the development of lioligand binding techniques which brought about the st quantification of adenosine receptors in a variety of sues. Radioligands for A_1 adenosine receptors which

---

*GLS is an Established Investigator of the American rt Association.

preferentially bind to A_1 compared to A_2 receptors have far outdistanced those which are putatively selective for A_2 adenosine receptors (see Jacobson et al. 1987). In addition, radioligand binding studies have begun to elucidate mechanisms responsible for the physiologic and pathologic regulation of A_1 adenosine receptors (Stiles 1986). The most recent and third genre of advances has been the introduction of agonist and antagonist photoaffinity probes to study the protein structure of the A_1 adenosine receptors.

II. STRUCTURE OF THE A_1 ADENOSINE RECEPTOR

The elucidation of the structure of membrane bound receptors has become an important goal in unraveling the mechanisms of neurotransmitter and hormone action at the molecular level (Stiles et al. 1984). Until 1985 there was virtually nothing known concerning the structure of A_1 adenosine receptors. In that year, three separate groups published reports on the synthesis and utilization of agonist A_1 adenosine receptor photoaffinity probes (Stiles et al. 1985, Klotz et al. 1985, Choca et al. 1985). Stiles et al. (1985) reported that the A_1 adenosine receptor from both rat brain and fat resides on a M_r 38,000 protein. Figure 1 demonstrates that the A_1 receptor binding subunit from brain demonstrates the appropriate pharmacological specificity and that consistent with the photoaffinity probe being an agonist, binding (labelling) is significantly decreased by guanine nucleotides (Gpp(NH)p). High concentrations (10^{-4}M) of Gpp(NH)p totally ablate labelling (not shown).

Very recently in collaboration with Dr. Ken Jacobson (Stiles & Jacobson 1987), we have synthesized a high affinity antagonist radioligand/photoaffinity crosslinking probe to label A_1 adenosine receptors. Figure 2 demonstrates the labelled binding subunit from bovine brain. It is of interest that the A_1 receptor from bovine brain is slightly atypical in that the agonist potency series in this receptor is R-PIA > S-PIA > NECA. As can be seen, [^{125}I]PAPAXAC being an antagonist probe labelling is unaffected by guanine nucleotides (Gpp(NH)p) in contradistinction to [^{125}I]APNEA labelling. Compare Figure 1 with Figure 2.

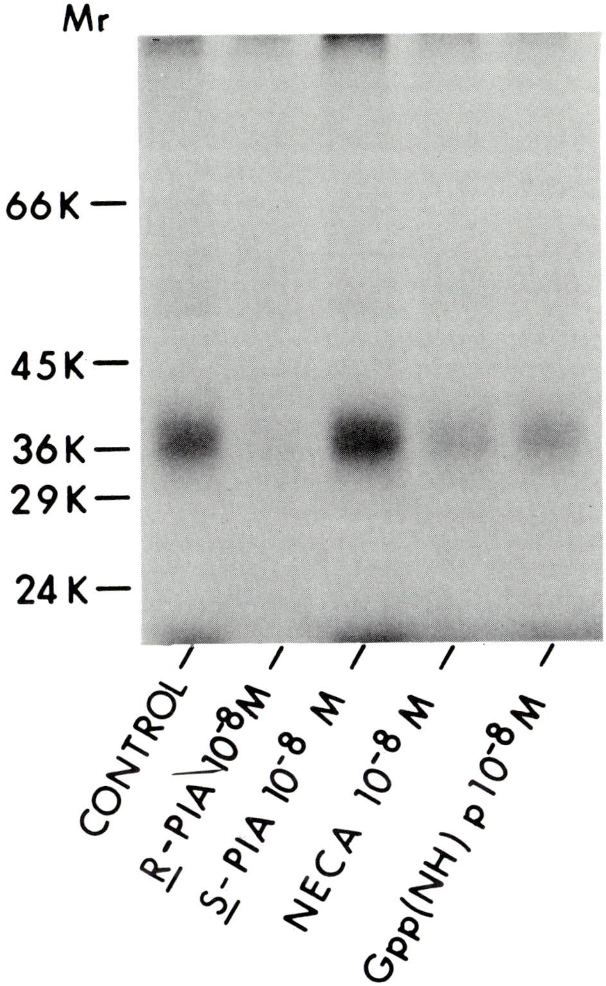

FIGURE 1. Photoaffinity cross-linking of [^{125}I]-APNEA into rat cerebral cortex membranes. Cerebral cortex membranes were photoaffinity cross-linked with [^{125}I]-APNEA alone (control) or in the presence of the indicated concentration of competing ligand as described in Stiles et al. (1985). The samples were then solubilized and electrophoresed on a 10% acrylamide gel. The relative molecular weight (M_r) scale was calibrated using iodinated protein standards.

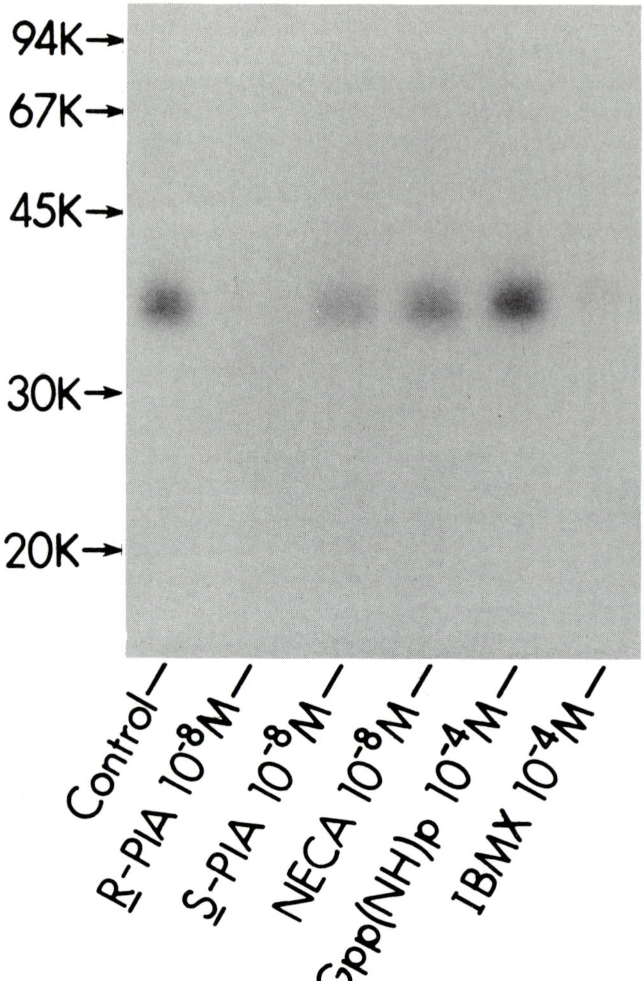

FIGURE 2. Photoaffinity crosslinking of [^{125}I]PAPAXAC into bovine cerebral cortex membranes. Cerebral cortex membranes were photoaffinity crosslinked with [^{125}I]PAPAXAC alone (control) or in the presence of the indicated concentration of competing ligands. The samples were then solubilized and electrophoresed on a 12% acrylamide gel. The relative molecular weight (M_r scale) was calibrated using iodinated protein standards.

This antagonist probe has several advantages over the agonist probes previously described in that binding is not dependent on receptor-guanine nucleotide regulatory protein (G protein) interactions and binding is unaffected by guanine nucleotides or divalent ions.

To study the more detailed structure of the A_1 adenosine receptor (A_1AR) and to compare the A_1AR's from different tissues, we have utilized the techniques of peptide mapping and endo- and exoglycosidase treatments of [^{125}I]APNEA-labelled receptor-binding subunits (Stiles 1986B). We have found that by peptide mapping techniques the A_1AR from rat brain and fat are similar both in terms of overall molecular weight (M_r 38,000) and by peptide mapping using a variety of proteolytic enzymes, such as papain and staph. aureus V8 proteinase. This technique utilizes two dimensional gel electrophoresis to directly compare the partial peptide maps of two proteins and thus allows one to determine if the proteins have structural similarities (homology) or if they are distinct proteins. Using this technique it was demonstrated that the peptide fragments generated from each of the two A_1 receptor binding subunits from brain and fat are equivalent (Stiles 1986B). This suggests that the primary structure of the A_1AR may not be very dissimilar between the two tissues.

Membrane receptors are all thought to be glycoproteins. To compare the carbohydrate nature of the A_1AR's from brain and fat, we subjected [^{125}I]APNEA labelled receptors to exoglycosidase and endoglycosidase treatments. Following neuramindase treatment, the A_1AR from both tissues migrate as M_r 36,000 protein an apparent shift from M_r 38,000 for untreated receptors. This suggests both receptors contain carbohydrate chains with terminal sialic acid (Stiles 1986B). Both [^{125}I]APNEA labelled A_1AR's were insensitive to α-mannosidase treatment suggesting that the A_1AR may not contain high mannose type chains which contain terminal mannose residues. Endoglycosidase F removes both N-linked high mannose and complex-type carbohydrate chains from proteins (Stiles 1986B). Following treatment with endoglycosidase F, both [^{125}I]APNEA labelled A_1AR migrate as M_r 32,000 proteins. This suggests that the A_1AR from these two tissues are glycoproteins which contain complex type carbohydrate chains and that following deglycosylation the receptors migrate as M_r 32,000 proteins. This suggests that the A_1AR's from rat fat and brain are identical in terms of their protein and carbohydrate structure.

Further insight into the biochemical nature of A_1AR's will require the purification and eventual molecular cloning of these receptors.

III. RECEPTOR REGULATION

Over the last decade it has become obvious that membrane bound receptors are not static entities but are dramatically regulated by a wide variety of pathophysiological conditions. Adenosine receptors are certainly no exception to this general phenomenon (see Stiles 1986 for a general review). There are now known to be a wide variety of drugs and physiological conditions which can alter the manner in which adenosine initiates the transmembrane signal to activate or deactivate its effector. For example, following methylxanthine treatment the A_1AR appears to be sensitized to the effects of adenosine analogs (Green & Stiles 1986). In addition, changes in thyroid hormone status, pregnancy/lactation or aging all appear to alter the sensitivity of tissues to adenosine or its analogs.

The phenomenon of desensitization is a ubiquitous adaptive mechanism utilized by all living creatures. The phenomenon is most commonly observed as a loss of responsiveness to a drug following prolonged or repeated exposure to that drug. In receptor mediated systems two types of refractoriness have been described - "homologous" and "heterologous" (Sibley & Lefkowitz 1985). The former refers to agonist-specific refractoriness, whereas, the latter connotes refractoriness to additional hormones or effectors.

Parsons & Stiles (1987) have recently studied the phenomenon of desensitization in the A_1AR - adenylate cyclase system of rat adipocytes. Rats were infused subcutaneously with R-PIA (75 nmol/kg/hr) or vehicle for six days and adipocyte membranes prepared. Fat membrane adenylate cycalse activity was then compared between control and agonist treated animals. Basal, isoproterenol stimulated, as well as, sodium fluoride and forskolin stimulated adenylate cyclase activity were all enhanced by approximately 100% in membranes from treated compared to control animals. Since each of these effectors stimulate adenylate cyclase activity by different pathways - isoproterenol through β receptors, NaF via G_s and forskolin principally via the catalytic unit, the data suggests that multiple components or a common converging component of the adenylate cyclase system must be altered following chronic exposure to R-PIA.

The ability of R-PIA to mediate inhibition of adenylate cyclase in adipocyte membranes was markedly attenuated in "treated" membranes. Figure 3 compares the R-PIA dose response curve for inhibition of forskolin-stimulated adenylate cyclase activity.

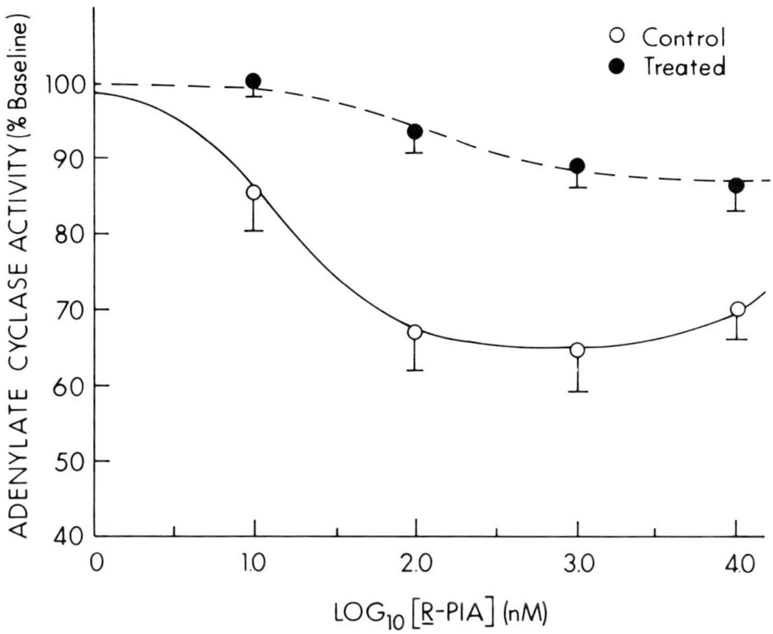

FIGURE 3. Inhibition of forskolin-stimulated adenylate cyclase activity in membranes from rat adipocytes. Membranes were prepared from control and R PIA-treated rats, and adenylate cyclase activities were measured as described (Parsons & Stiles 1987). Data are presented as mean ± S.E. for five experiments. Forskolin (10^{-5}M)-stimulated adenylate cyclase activities were 235 ± 55 and 336 ± 39 pmol/mg of protein/min in membranes from control and treated rats, respectively. R-PIA was added at the concentrations indicated on the abscissa. Adenylate cyclase inhibition was significantly ($p < 0.001$) attenuated in membranes from treated animals.

Attenuation of A_1 adenosine receptor-mediated inhibition was noted regardless of whether basal, forskolin or isoproterenol stimulated activity was used as a baseline.

Rat adipocyte membranes also contain PGE_1 receptors which act via the G_i protein to inhibit adenylate cyclase activity. Following R-PIA treatment it was found that there was a 50% attentuation of PGE_1's ability to inhibit adenylate cyclase. This is consistent with a heterologous type of desensitization.

Using the agonist radioligand [^{125}I]HPIA, saturation curve analysis revealed a 32% decrease in A_1AR in treated compared to control membranes (1176 ± 332 fmol/protein vs 1732 ± 253 mg). Since [^{125}I]HPIA binds only to the high affinity state of the receptor at the concentrations utilized, this change could represent a loss of high affinity state or a true loss of A_1AR's from the membranes. Further analysis using high affinity antagonists will be necessary to differentiate these two possibilities. These data are sufficient, however, to document that there are alterations in receptor-effector coupling following chronic agonist exposure. To determine if there is a change in receptor structure, the A_1AR from control and treated membranes were photoaffinity labelled with [^{125}I] AZPNEA (Stiles et al. 1986). Following SDS-PAGE/autoradiography, no alteration in receptor structure was detected following desensitization.

To elucidate a mechanism which could explain the apparent sensitization of stimulatory effectors, as well as the diminished sensitivity to inhibitory effectors, we probed the quantity and functionality of the guanine nucleotide regulatory proteins G_s and G_i.

Bacterial toxin labelling techniques have greatly facilitated the ability to quantitate the α subunits of G_s and G_i, the mediators of stimulation and inhibition of adenylate cyclase, respectively. Cholera toxin promotes the ADP ribosylation of α_s and pertussis toxin α_i.

Under appropriate conditions, the amount of [^{32}P]NAD incorporated into these peptides is believed to reflect the amount of α_s and α_i in membranes. Using cholera and pertussis toxins, α_s and α_i were quantitated in membranes from control and treated rats. Figure 4 demonstrates a representative experiment in which α_i (M_r 42,000) and α_s (M_r 42,000 and 47,000) are quantitated.

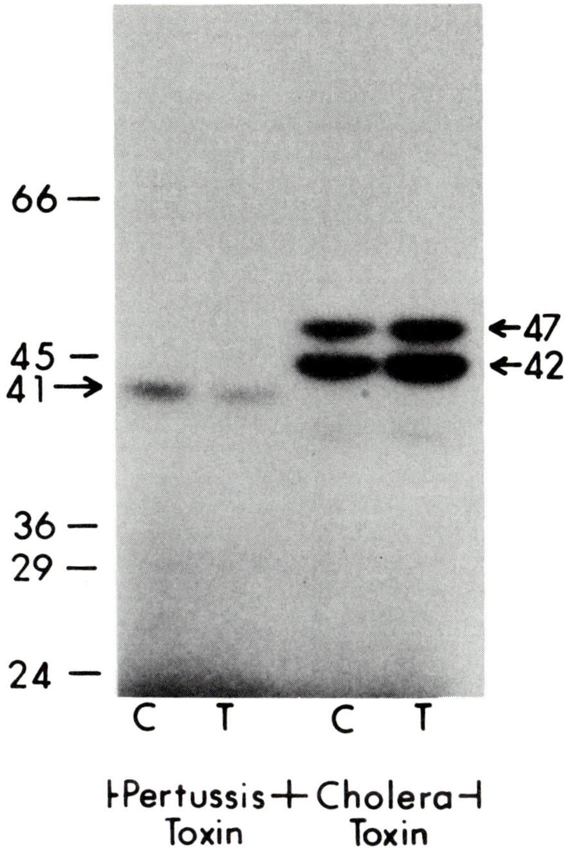

FIGURE 4. Pertussis toxin- and cholera toxin-
ıediated [^{32}P]NAD labelling of control and R-PIA-
desensitized' fat membranes. Fat membranes derived from
ne control (C) and one R-PIA-treated (T) rat were divided
nto aliquots and subjected to [^{32}P]NAD labelling using
ertussis toxin and cholera toxin as described in Parsons
Stiles (1987). The samples were then subjected to SDS-
AGE and autoradiography. The M_r 41,000 protein in the
irst two lanes represents the α_i subunit of G_i. The
$_r$ 47,000 and 42,000 proteins in the third and fourth
anes are the two major proteins representing the α_s sub-
ıit of G_s. There was a 37% decrease in M_r 41,000 sub-
trate in membranes from treated animals and a 44%
ıcrease in the α_s substrates in treated compared to
ontrol membranes.

From these types of experiments α_s is found to be increased 44% and α_i is decreased by 37% in "treated" compared to "control" membranes. These alterations in the coupling proteins, G_s and G_i, could explain all the phenomenon observed in adenylate cyclase assays.

To put these findings in perspective, we can state that exposure of rat adipocytes to R-PIA leads to a diminished capacity of the A_1AR system to inhibit adenylate cyclase activity concomitant with an enhanced responsiveness to stimulatory agents. The common converging pathway for stimulation of cyclase is the G_s protein (α_s) while for the inhibitory arm of the pathway the G_i (α_i) protein is the final common pathway. Thus, a reciprocal regulation of G_s and G_i would be expected to produce enhanced activation of adenylate cyclase and a blunted inhibition by inhibitory agents. This study raises many more questions than it answers. For example, how does an inhibitory agonist modulate G_s and G_i levels. Are A_1AR's regulated quantitatively or is just high affinity coupling modulated. Further studies will be required to answer these questions.

SUMMARY

Much new information on the structure and regulation of A_1 adenosine receptors has become available in the last few years. This has been a consequence of recent technological advances in the tools for studying this receptor system. The next few years should provide ever expanding vistas in terms of structure-function relationships and receptor regulation at the biochemical and molecular biological level.

REFERENCES

Choca, J.L., Kwatra, M.M., Hosey, M.M. & Green R.D., 1985, Biochemical and Biophysical Research Communications, 131, 115.

Green, R.M. & Stiles, G.L., 1986, Journal of Clinical Investigation, 77, 222.

Jacobson, K.A., Ukena, D., Padgett, W., Kirk, K.L. & Daly, J.W., 1987, Biochemical Pharmacology, 36, 1697.

Klotz, K.N., Cristalli, G., Grifantino, M., Victari, S. & Lohse, M.I., 1985, Journal of Biological Chemistry, 260, 14659.

Parsons, W.J. & Stiles, G.L., 1987, Journal of Biological Chemistry, 262, 841.

Sibley, D. & Lefkowtiz, R.J., 1985, Nature, 317, 124.

Stiles, G.L., 1986, Trends in Pharmacological Sciences, 7, 486.

Stiles, G.L., 1986B, Journal of Biological Chemistry, 261, 10839.

Stiles, G.L., Caron, M.G. & Lefkowitz, R.J., 1984, Physiological Reviews, 64, 661.

Stiles, G.L, Daly, D.T. & Olsson, R.A., 1985, Journal of Biological Chemistry, 260, 10806.

Stiles, G.L., Daly, D.T. & Olsson, R.A., 1986, Journal of Neurochemistry, 260, 14659.

Stiles, G.L. & Jacobson, K.A., 1987, Molecular Pharmacology, 32, 184.

RECEPTORS FOR ADENINE NUCLEOTIDES

Studies on the P_2-purinoceptor using adenine nucleotide analogues

N.J. CUSACK, L.A. WELFORD & S.M.O. HOURANI*

Department of Pharmacology, King's College London, Strand, London WC2R 2LS, U.K., and *Department of Biochemistry, University of Surrey, Guildford GU2 5XH, U.K.

INTRODUCTION

Adenine nucleotides have pharmacological effects on a wide variety of tissues, by acting on receptors which have been called P_2-purinoceptors to distinguish them from P_1-purinoceptors which recognise adenosine (Burnstock 1978[1]). In particular, ATP and ADP cause contraction or relaxation of many smooth muscle preparations, and ATP has been proposed as a non-adrenergic, non-cholinergic neurotransmitter or co-transmitter in the autonomic nervous system (Burnstock 1986). For example, ATP contracts the guinea-pig urinary bladder and vas deferens, and the longitudinal muscle of the rat portal vein, and the rat anococcygeus, but relaxes the guinea-pig taenia coli and aorta. The relaxant effect on blood vessels depends on the presence of an intact endothelium, which releases endothelium-derived relaxing factor (EDRF), whereas in the absence of the endothelium, ATP causes constriction of blood vessels, presumably by a direct effect on the underlying smooth muscle (Furchgott 1984).

As well as having pharmacological effects, adenine nucleotides are rapidly and sequentially dephosphorylated by ectonucleotidases present on the surface of smooth muscle, to give ultimately adenosine (Welford 1986; 1987). Since adenosine in general relaxes smooth muscle, this dephosphorylation may influence the observed effects of ATP, particularly in tissues where ATP causes contraction. This should be borne in mind when considering the reported effects of adenine nucleotides and their analogues (for further discussion see Cusack et al. this volume).

The stucture-activity relationships for analogues of ATP are not identical for these two effects of ATP, contraction and relaxation of smooth muscle, and this has led to the suggestion that P_2-purinoceptors can be subdivided into two classes, P_{2X} causing contraction and P_{2Y} causing relaxation (Burnstock & Kennedy 1985; Gordon 1986). In this chapter, we will consider evidence in favour of this suggestion, which has come from the study of the structure-activity relationships of adenine nucleotide analogues in two representative tissues, the guinea-pig taenia coli (P_{2Y}) and the guinea-pig urinary bladder (P_{2X}). The effects of ATP analogues modified in the adenine base, the ribose sugar and the triphosphate chain will be discussed, and in addition some new synthetic analogues which are specific agonists for these putative receptor subtypes will be described.

STRUCTURE-ACTIVITY RELATIONSHIPS

Adenine-modified analogues of ATP

As well as ATP, the other naturally occurring ribonucleoside triphosphates GTP, UTP and CTP also induce relaxation of the guinea-pig taenia coli and contraction of the urinary bladder, and are only a little less potent than ATP for both effects (Lucaksco & Krell 1982; Cusack & Beese unpublished). Although ATP and ADP are equipotent, the other naturally occurring diphosphates GDP, UDP and CDP are much less active at inducing relaxation or contraction (Lucaksco & Krell 1982; Cusack & Beese unpublished), while AMP is 30 times less potent than ATP at inducing relaxation and is completely inactive in the urinary bladder (Satchell & Maguire 1975; Burnstock et al. 1984).

Gough et al. (1973), following up their work on ADP receptors on platelets, where 2-substituted analogues of ADP are up to 10 times as active as ADP itself, were the first to show that the corresponding ATP analogues, 2-chloro-ATP and 2-methylthio-ATP are 3 and 50 fold more potent than ATP at relaxing the taenia coli. This effect of substituents on the 2-position was confirmed by Cusack & Planker (1979) who showed that the photoaffinity analogue 2-azido-ATP is 20 times as potent as ATP, and by Burnstock et al. (1983) who found that 2-chloro-ATP is 30 times and 2-methylthio-ATP 200 times more potent than ATP. Differences in potencies by Gough et al. (1973) from those of Burnstock et al. (1983) probably reflect differences in methodology. They used tissues that had a naturally high tone, whereas Burnstock et al. (1983) used tissues that had first been contracted by carbachol. In contrast to the taenia coli, the 2-substituted analogues are no more potent than ATP at contracting the urinary bladder (Burnstock et al. 1983). The 8-substituted analogue 8-bromo-ATP is about as potent as ATP in both tissue preparations (Satchell & Maguire 1983; Welford et at., 1987), while N^6-phenyl-ATP is inactive on the taenia coli and the urinary bladder (Welford at al., 1987; and unpublished).

Ribose-modified analogues of ATP

Replacement of the naturally occurring D-ribose sugar of adenine nucleotides by the unnatural L-ribose generates their L-enantiomers, and the relative potencies of these pairs of enantiomers enables the stereoselectivity of P_2-purinoceptors to be determined (Cusack & Planker 1979). ATP is 3 times more potent than L-ATP, and ADP is only 6 times as potent as L-ADP, showing that the inhibitory P_{2Y}-purinoceptor on the taenia coli has only poor stereoselectivity for the naturally occurring enantiomers. However, 2-substitution greatly improves this stereoselectivity due to the increased potency of the natural enantiomer over that of the unnatural enantiomer, and 2-azido-ATP and 2-azido-ADP are both 125 times more potent than 2-azido-L-ATP and 2-azido-L-ADP, respectively (Cusack & Planker 1979). The poor stereoselectivity on the P_{2Y}-purinoceptor and the great improvement following 2-substitution is confirmed by Burnstock et al. (1983), who found that 2-chloro-ATP is 54 times more potent than 2-chloro-L-ATP, and 2-methylthio-ATP is more than 700 times as potent as 2-methylthio-L-ATP. In contrast, the excitatory P_{2X}-purinoceptor on the urinary bladder shows no stereoselectivity towards the enantiomers of ATP, ADP, 2-chloro-ATP or 2-methylthio-ATP, and indeed the unnatural L-enantiomers are at some concentrations more potent than the natural enantiomers (Burnstock et al. 1983; Welford et al. 1987). Unlike AMP, L-AMP is inactive on the taenia coli (Cusack & Planker 1979) and both enantiomers are inactive on the urinary bladder (Welford et al. 1987). The P_1-purinoceptor for adenosine on both the taenia coli and the urinary bladder shows absolute stereoselectivity for the natural enantiomer as L-adenosine is inactive in both tissues (Cusack & Planker 1979; Cusack & Hourani 1984). Other analogues of ATP with modifications to the ribose sugar that have been tested are ara-ATP, where the 2'-hydroxyl has been inverted, and 2-deoxy-ATP where the 2'-hydroxyl has been replaced by hydrogen, and these analogues have less potency than ATP at relaxing the taenia coli (Maguire & Satchell 1983), but their activity on the urinary bladder is not reported.

Phosphate-modified analogues of ATP

In view of the obvious importance of the triphosphate chain to the binding of ATP to P_2-purinoceptors, the effects of several types of modifications have been examined, which include replacement of ionised phosphate oxygen atoms by other atoms or groups, and replacements of the interphosphate oxygens by other linkages.

Phosphorothioates. Replacement of an ionised oxygen by by ionised sulphur generates phosphorothioate analogues of adenine nucleotides (Eckstein & Goody 1976). If the sulphur is attached to a non-terminal phosphate, then that phosphorus is rendered chiral so that the analogues exist as pairs of R_p and S_p diastereoisomers, which can provide furthur information about the stereoselectivity of the P_2-purinoceptor. All of the phosphorothioate analogues of ATP, ADP and AMP are at least as potent as the parent nucleotides at relaxing the taenia coli and contracting the urinary bladder (Burnstock et al. 1984). Stereoselectivity by the taenia coli is exhibited only towards the diastereoisomers of ATP -α-S and ADP-α-S, analogues having the sulphur joined to the innermost phosphate, and the R_p isomers are the more potent analogues at inducing relaxation. In addition, R_p-ATP-α-S and Rp-ADP-α-S are 50 times and 9 times more potent than ATP and ADP respectively, though a later study (Stone 1985) does not confirm this. The diastereoisomers of ATP-β-S, where the sulphur is attached to the middle phosphate, as well as ATP-γ-S, ADP-β-S and AMPS where the sulphur is attached to the terminal phosphate, are about as potent as ATP, ADP and AMP, respectively. In contrast to the responses of the taenia coli, the urinary bladder fails to exhibit stereoselectivity for any of the diastereoisomers of ATP-α-S or of ATP-β-S, and it is ATP-β-S instead of ATP-α-S that is much more potent than ATP, and that unlike ATP, ADP or the rest of the phosphorothioate analogues, has a sigmoid log. dose-response curve. ATP-γ-S, ATP-α- S, ADP-β-S, ADP-α- S and AMPS are all about equipotent with ATP at contracting the urinary bladder (Burnstock et al. 1984). The activity of AMPS provides some evidence that the reason that AMP itself is inactive is because it fails to stimulate adequately the P_{2X}-purinoceptor, rather than by acting at an adenosine receptor either in its own right or as a consequence of prior dephosphorylation.

Methylenephosphonates. Replacement of one of the bridging oxygens joined to phosphate by a methylene group generates a methylenephosphonate linkage which is itself completely stable to chemical or enzymic hydrolysis (Myers et al. 1963). Incorporation of this linkage into ATP can occur between the ribose sugar and the innermost phosphate to provide homo-ATP (Jones & Moffatt 1968), the innermost and middle phosphates to provide α,β-methylene-ATP (AMPCPP) and the middle and terminal phosphates to provide β,γ-methylene-ATP (AMP-PCP). The potencies of these compounds as relaxants of the taenia coli depend on the point of insertion of the methylene linkage into the triphosphate chain. Homo-ATP is some 70 times more potent than ATP, α,β-methylene-ATP has a similar potency to ATP, while β,γ-methylene-ATP has only about 0.2 times the potency of ATP (Satchell & Maguire 1975; Maguire & Satchell 1979). Comparison of potencies of these analogues is complicated by the shapes of their log. dose response curves, which in the high tone tissues used are much steeper than that of ATP, and give supramaximal responses up to 1.4 times that of ATP. On tissues that have firstly been contracted by carbachol, the log. dose-response curve to β,γ-methylene-ATP is parallel to that of ATP, and even though this methodology is different a similar value for the relative potency to ATP is obtained (Cusack et al. 1987). The effect of 2-substituents is to enhance the potency, and 2-chloro-β,γ-methylene-ATP and 2-methylthio-β,γ-methylene-ATP are 3.2 and 2.5 times, respectively, as potent as β,γ-methylene-ATP, though both analogues are still not as potent as ATP itself (Cusack et al. 1987; and unpublished)

The potencies of the methylenephosphonate analogues on the urinary bladder also depends on the position of the methylene linkage in the triphosphate chain, but the order is the reverse of that for relaxation. Homo-ATP is no more effective than ATP (Welford et al. 1987), while α,β-methylene-ATP (Kasakov & Burnstock 1983; Welford at al. 1987) and β,γ-methylene-ATP (Lucacsko & Krell 1982; Cusack & Hourani 1984) are much more effective at inducing contraction of the bladder. Unlike the taenia coli, 2-substitution of the adenine ring does not lead to analogues having enhanced potencies, and 2-methylthio-β,γ-methylene-ATP and 2-chloro-β,γ-methylene-ATP are no more potent than β,γ-methylene-ATP (Cusack et al. 1987; and unpublished).

Isopolar phosphonates. Although replacement of an interphosphate oxygen by a methylene group provides a stable linkage that is approximately isosteric with the parent diphosphate moiety, the resultant methylenediphosphonate does not mimic adequately the electronic nature of the triphosphate chain of ATP. Thus replacement of the electronegative oxygen by the electropositive methylene group means that β,γ-methylene ATP is not as fully ionised as ATP at physiological pH (Myers et at . 1963), and this has been considered as an explanation for the relative lack of potency of this analogue as a relaxant of the taenia coli (Satchell & Maguire 1975). To overcome this deficiency, dihalomethylene gruops have been used instead of methylene to restore the electronegative character of this interphosphate linkage so as to provide isopolar isosteric phosphonate analogues which are fully ionised at physiological pH (Blackburn et al. 1981). Two of these analogues, β,γ-dichloromethylene-ATP and β,γ-difluoromethylene-ATP are 1.4 and 2.6 times, respectively, more potent than β,γ-methylene-ATP at relaxing the taenia coli, an order of potency that refects the order of electronegativity of the analogues, but both are still less potent than ATP. However, 2-substituents greatly enhance their potencies, and 2-methylthio-β,γ-dichloromethylene-ATP is 11 times and 2-methylthio-β,γ-difluoromethylene-ATP 130 times, respectively, as potent as β,γ-methylene-ATP (Cusack et al. 1987). Indeed 2-methylthio-β,γ- difluoromethylene-ATP is twice as potent as ATP itself, but it is not as potent as 2-methylthio-ATP, which illustrates that electronegativity is not the only factor determining pharmacological potencies as relaxants. Dihalomethylene groups, unlike the oxygen atoms they replace, do not have any lone pairs of electrons that may be important for binding to the P_2-purinoceptor. β,γ-Imido-ATP, an isopolar ATP analogue in which the oxygen has been replaced by a nitrogen, has an imidodiphosphate linkage which, though not always stable to chemical or enzymic hydrolysis, is sterically very similar to the diphosphate group and addition does possess a lone pair of electrons (Yount et al. 1971). However, β,γ-imido-ATP is no more potent than β,γ-methylene-ATP at relaxing the taenia coli (Maguire & Satchell 1979).

In contrast, the isopolar analogues are no more potent than β,γ-methylene-ATP at contracting the urinary bladder, and the order of potency does not reflect their order of electronegativity, as β,γ-difluoromethylene-ATP is about as potent as β,γ-methylene-ATP and β,γ-dichloromethylene-ATP is less effective. 2-Substitution does not generate more potent analogues, and the order of potency again does not refect the order of electronegativity (Cusack et al. 1987). The comparitive lack of potencies of the dichloromethylene analogues may be due to some distortion of the triphosphonate chain necessary to accomodate the bulky chlorine groups. The isopolar imido analogue, β,γ-imido-ATP although much more potent than ATP, was no more potent than β,γ-methylene-ATP, and so lone pairs of electrons seem of little importance for activity of ATP analogues on the urinary bladder (Welford et al. 1987). It should be emphasised that as none of these analogues are degraded by ectonucleotidases on smooth muscle, their differences in potencies are uncomplicated by the effects of any differences in their rates of dephosphorylation (see Cusack et al. this volume).

SPECIFIC AGONISTS FOR EXCITATORY P_{2X}-PURINOCEPTORS

The structure-activity requirements for the excitatory P_{2X}-purinoceptor on the guinea-pig urinary bladder seem to be less rigid than those for the inhibitory P_{2Y}-purinoceptor on the guinea-pig taenia coli (figure 1). In particular, the contractile response is unaffected by replacement of ribose by its L-enantiomer as in L-ATP, and placing a methylene linkage between the middle and terminal phosphates as in β,γ-methylene-ATP leads to greatly enhanced potency. In contrast, each of these modifications leads to analogues having reduced potencies as relaxants. It seemed therefore that by combining both modifications in the same ATP analogue, it might be possible to generate a specific agonist for the P_2-purinoceptor mediating contraction. We therefore synthesised β,γ-methylene-L-ATP (L-AMP-PCP), and found that this analogue is the most potent known ATP agonist at the P_{2X}-purinoceptor mediating contraction of the urinary bladder (Cusack & Hourani 1984), while having no agonist or antagonist action at the P_{2Y}-purinoceptor mediating relaxation of the taenia coli (Hourani et al. 1985). Since specific competitive inhibitors of the actions of ATP are not known, desensitisation was used to ensure that β,γ-methylene-L-ATP induces contraction by acting at the same receptor as ATP. β,γ-Methylene-L-ATP causes a rapid desensitisation of the urinary bladder to nerve stimulation, which is also cross-desensitised to ATP but not to histamine, showing that the action of the analogue is specific to the P_{2X}-purinoceptor on this tissue (Hourani 1984). Although β,γ-methylene-L-ATP is not dephosphorylated, its possible breakdown product is L-adenosine, which unlike adenosine is inactive on the bladder and so cannot contribute any nonspecific inhibition. In particular, adenosine inhibits the response to nerve stimulation whereas L-adenosine does not (Cusack & hourani 1984). β,γ-Methylene-L-ATP has been used to identify excitatory P_{2X}-purinoceptors in a variety of isolated smooth muscle preparations where ATP is active. In particular, β,γ-methylene-L-ATP is much more potent than ATP at contracting the guinea-pig vas deferens, the rat portal vein and anococcygeus, but has virtually no action on tissues such as the guinea-pig field-stimulated ileum, where ATP causes relaxation. Unlike ATP and β,γ-methylene-ATP, which relax the guinea-pig thoracic aorta, β,γ-methylene-L-ATP actually contracts this tissue, due to a selective direct action on the vascular smooth muscle rather than by an indirect action via an endothelium-dependent mechanism (Hourani et al. 1986). The action of β,γ-methylene-L-ATP on the rat hepatic portal vein was later confirmed by Reilly & Burnstock (1987), who showed that the electrophysiological effects of β,γ-methylene-L-ATP were similar to those of ATP itself. A family of specific agonists for the excitatory P_{2X}-purinoceptor has since been synthesised based on the β,γ-methylene-L-ATP molecule. The isopolar versions, β,γ-dichloromethylene-L-ATP and β,γ-difluoromethylene-L-ATP, are both inactive on the taenia coli, but are potent agonists on the urinary bladder (Cusack et al. 1987), and the 2-substituted analogues 2-chloro-β,γ-methylene-L-ATP and 2-methylthio-β,γ-methylene-L-ATP are also inactive on the taenia coli, but very potent on the urinary bladder.

A SPECIFIC AGONIST FOR INHIBITORY P_{2Y}-PURINOCEPTORS

Because the structure-activity requirements of the excitatory P_{2X}-purinoceptor are so much less demanding than those of the inhibitory P_{2Y}-purinoceptor, it seemed that any ATP analogue that is active on the taenia coli would also be active on the urinary bladder. The structure-activity relationships of ADP and AMP analogues, however, appear to be rather more constrained. Thus ADP contracts the urinary bladder but the other naturally occurring diphosphates GDP, CDP and UDP do not (Lucacsko & Krell 1982), and ADP relaxes the taenia coli but GDP, CDP and UDP do not (Cusack & Beese, unpublished). AMP relaxes the taenia coli but is inactive on the urinary bladder, while its phosphorothioate analogue, AMPS is as potent as ATP at inducing relaxation (Burnstock et al. 1984). The ADP analogue, adenosine 5'-(2-fluorodiphosphate) (ADP-β-F), is unusual in that it has the diphosphate chain of ADP but has only the same number of negative charges as AMP.

ADP-β-F induces relaxation of the taenia coli, but has no action on the urinary bladder either as agonist or as antagonist (Hourani et al. 1987). Its lack of potency cannot be due to rapid degradation to adenosine, as the fluorine atom on the terminal phosphate confers some resistance to dephosphorylation (Welford et al. 1986; Hourani et at. 1987).

FIGURE 1A. MODIFICATIONS OF THE STRUCTURE OF ATP WHICH RESULT IN ENHANCED POTENCY AT INHIBITORY P_{2Y}-PURINOCEPTORS

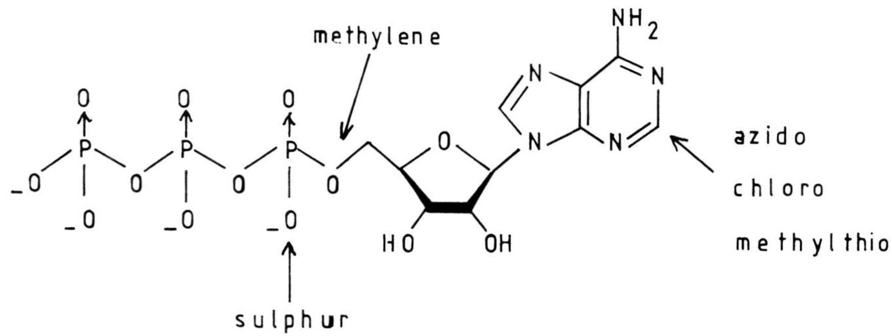

FIGURE 1B. MODIFICATIONS OF THE STRUCTURE OF ATP WHICH RESULT IN ENHANCED POTENCY AT EXCITATORY P_{2X}-PURINOCEPTORS

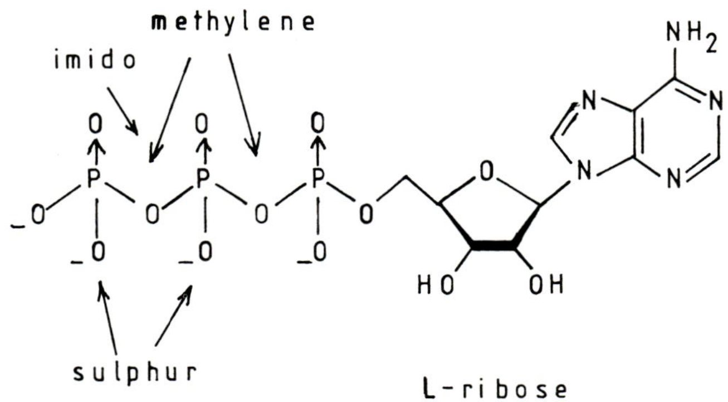

ADP-β-F, like ATP, relaxes the guinea-pig aorta, but has no action on the guinea-pig vas deferens where ATP induces contraction, and ADP-β-F is therefore a specific agonist for the inhibitory P_{2Y}-purinoceptor mediating relaxation (Hourani et al. 1987)

CONCLUSIONS

The suggestion by Burnstock & Kennedy (1985) that P_2-purinoceptors could be divided into two subclasses is based on structure-activity relationships for ATP analogues, some of which, notably 2-methylthio-ATP, are dephosphorylated as rapidly as ATP itself (Welford et al. 1986; 1987). Our more recent work, using analogues which are not degraded confirms that there are differences in structure-activity between the P_2-purinoceptors on the taenia coli and the urinary bladder, and that two receptor subtypes could exist. The discovery that β,γ-methylene-L-ATP and its derivatives are specific agonists for the putative P_{2X}-purinoceptor, and that ADP-β-F is a specific agonist for the P_{2Y}-purinoceptor (Cusack et al. 1987; Cusack & Hourani 1984; Hourani et al. 1985; 1986; 1987) strengthens this suggestion further. However, as all of these analogues are agonists, affinity for the receptor is not the only factor contributing to the observed potencies, and these differences could instead conceivably be explained by differences in efficacy or in the numbers of receptors in each tissue (Kenakin 1984). Conclusive proof of the existence of two receptor types must await the discovery of convincing reversible selective competitive antagonists.

REFERENCES

Blackburn, G.M., Kent, D.E. & Kolkmann, F., 1981, Journal of the Chemical Society Chemical Communications, 1188.

Burnstock, G., 1986, Acta Physiologica Scandinavia, 126, 67.

Burnstock, G., 1978 a basis for distinguishing two types of purinergic receptor. In Cell Membrane Receptors for Drugs and Hormones: a Multidisciplinary Approach, edited by R.W. Straub and L. Bolis (Raven Press), p.107.

Burnstock, G., Cusack, N.J., Hills, J.M., MacKenzie, I. & Meghji, P., 1983, British Journal of Pharmacology, 79, 907.

Burnstock, G. & Kennedy, C., 1985, General Pharmacology, 16, 433.

Cusack, N.J. & Hourani, S.M.O., 1984, British Journal of Pharmacology, 82, 155.

Cusack, N.J., Hourani, S.M.O., Loizou, G.D. & Welford, L. A., 1987, British Journal of Pharmacology, 90, 791.
Cusack, N.J. & Planker, M., 1979, British Journal of Pharmacology, 67, 153.
Eckstein, F. & Goody, R.S., 1976, Biochemistry, 15, 1685.
Furchgott, R.F., 1984, Annual Reviews of Pharmacology and Toxicology, 24, 175.
Gordon, J.L., 1986, Biochemistry Journal, 233, 309.
Gough, G.R., Maguire, M.H. & Satchell, D.G., 1973, Journal of Medicinal Chemistry, 16, 1188.
Hourani, S.M.O., 1984, British Journal of Pharmacology, 82, 161.
Hourani, S.M.O., Cusack, N.J. & Welford, L.A., 1985, European Journal of Pharmacology, 108, 197.
Hourani, S.M.O., Welford, L.A., Loizou, G.D. & Cusack, N.J., 1987, European Journal of Pharmacology, in the press.
Jones, G.H. & Moffatt, J.G., 1968, Journal of the American Chemical Society, 90, 5337.
Kasakov, L. & Burnstock, G., 1983, European Journal of Pharmacology, 86, 291.
Kenakin, T.P., 1984, Pharmacological Reviews, 36, 165.
Lucacsko, P. & Krell, R.D., 1982, Canadian Journal of Physiology and Pharmacology, 59, 1199.
Maguire, M.H. & Satchell, D.G., 1979 specificity of adenine nucleotide receptor sites: inhibition of the guinea-pig taenia coli by adenine nucleotide analogues. In Physiological and Regulatory Functions of Adenosine and Adenine Nucleotides, edited by H.P. Baer and G.I. Drummond (Raven Press), p. 33.
Myers, T.C., Nakamura, K. & Fleshner, J.W., 1963, Journal of the American Chemical Society, 85, 3292.
Reilly, W.M. & Burnstock, G., 1987, European Journal of Pharmacology, 138, 319.
Satchell, D.G. & Maguire, M.H., 1983 inhibitory purinergic receptors in visceral smooth muscle. In Physiology and Pharmacology of Adenosine Derivatives, edited by J.W. Daly, Y. Kuroda, J.W. Phyllis, H. Shimizu and M. Ui (Raven Press), p. 85.
Satchell, D.G. & Maguire, M.H., 1975, Journal of Pharmacology and Experimental Therapeutics, 195, 5430.
Stone, T.W., 1985, British Journal of Pharmacology, 84, 165.
Welford, L.A., Cusack, N.J. & Hourani, S.M.O., 1987, European Journal of Pharmacology, in the press.
Welford, L.A., Cusack, N.J. & Hourani, S.M.O., 1986, European Journal of Pharmacology, 129, 217.
Yount, R.G., Babcock, D., Ballantyne, W. & Ojala, D., 1971, Biochemistry, 10, 2484.

DIFFERENCES IN THE STRUCTURAL REQUIREMENTS FOR AGONIST PROPERTIES AT P_1 AND P_2 RECEPTORS IN SMOOTH MUSCLE.

David Satchell

Department of Zoology, University of Melbourne,
Parkville, Victoria 3052, Australia.

INTRODUCTION

A valuable insight into the mechanism of action of purine nucleosides and nucleotides in smooth muscle preparations can be obtained by studying the agonist properties of a series of structural analogues on the different receptor types. In the present experiments, the actions of agonists with modifications of the purine moiety and in particular to the ribofuranose ring are examined on mammalian vascular and gut preparations known to contain P_1 and P_2 receptors and P_2 receptor subtypes. Structural characteristics which promote agonist properties at the different sites are considered.

METHODS

Whole Vein Preparations

Male or female New Zealand white rabbits and male or female Sprague Dawley rats were killed by a blow to the head and exsanguination. Portal veins were excised and dissected free of surrounding connective tissue. Endothelial linings of vessels were removed by osmotic shock in cold distilled water and mechanical rubbing. Preparations were mounted vertically in organ baths and allowed to equilibrate for 60 min at a resting tension of 0.5 g for rabbit preparations and 0.2 g for rat preparations.

Arterial Strip Preparations

Femoral arteries were dissected from rabbits prepared as above. 2 cm lengths of spiral strips 0.5 mm wide, were cut from the artery and mounted vertically in organ baths at a tension of 0.1 g.

Taenia Coli Preparations

Guinea-pigs were killed by a blow to the head and exsanguination. Strips of taenia coli were dissected free of underlying caecum and mounted in organ baths at a resting tension of 1 g.

Nutrient Media

All preparations were mounted in a Kreb's solution of the following composition (millimolar-concentrations): Na Cl, 133; Ca Cl_2, 2.5; $MgSo_4$, 1.2; $NaHCO_3$, 16.3; NaH_2PO_4, 1.5; dextrose, 7.8; Kcl, 4.7 bubbled with a mixture of 95% oxygen - 5% carbon dioxide and maintained at 36°C.

Recording Techniques

Muscle activity was registered on a Grass polygraph model 7D and recorded isometrically with Grass FT03 transducers in guinea-pig taenia coli and rabbit portal vein and Gould-Statham UC2 transducers for the remaining preparations.

Adenosine and Adenine Nucleotides Analogues Used

ATP, adenosine, adenine-9-β-D-arabinofuranoside, 5'-triphosphate (ara-ATP), 8-bromo-adenosine-5'-triphosphate (8-bromo-ATP), 2'-deoxyadenosine 5'-triphosphate (2'-deoxy ATP), cordycepin triphosphate (3'-deoxy-ATP) and tetrodotoxin were obtained from the Sigma Chemical Company, St.Louis.

At the conclusion of organ bath experiments, tetrodotoxin (1 μg/ml) was added to the organ bath and the effects of agonists tested to rule out the possibility of nerve mediated responses.

RESULTS AND DISCUSSION

Fundamental Differences Between P_1 and P_2 Receptors

Burnstock (1978) distinguished two types of purinergic receptors: The P_1 purinoceptor had the order of potency: adenosine ⩾ AMP >ADP ⩾ ATP, while the P_2 purinoceptor had the reverse order. On guinea-pig taenia coli strips dose-response curves plotted for relaxations to ATP and ADP were almost superimposed as were the dose-response curves to AMP and adenosine. The ATP/ADP curves were displaced considerably to the left of the AMP/adenosine curves and the latter pair exhibited a distinctive sigmoidal shape not shown by the former (Maguire & Satchell 1981). Other fundamental differences in the receptor types were seen in tissues where ATP and adenosine caused opposite responses, moreover, in these circumstances ADP caused the same effect as ATP and AMP the same as adenosine. For example, in preparations of the detrusor muscle of both rat and guinea-pig, ATP and ADP caused a contraction while adenosine and AMP caused relaxations (Maguire & Satchell 1981; Burnstock & Brown 1981). The same type of result was obtained in preparations of chicken rectum (Burnstock & Brown 1981). In preparations of guinea-pig vas deferens. ATP and ADP caused contractions where AMP and adenosine were inactive (Holck & Marks 1978). In cat renal artery preparations ATP and ADP caused vasodilatation whereas AMP and adenosine caused vasoconstriction (Burnstock & Brown 1981). In guinea-pig tracheal strips where adenosine and AMP are active relaxants, the presence of receptors for ATP have been questioned. The metabolically stable methylene isosteres of both ATP and ADP were inactive. This suggests the P_2 receptors upon which ATP and ADP act are absent (Satchell, unpublished observations).

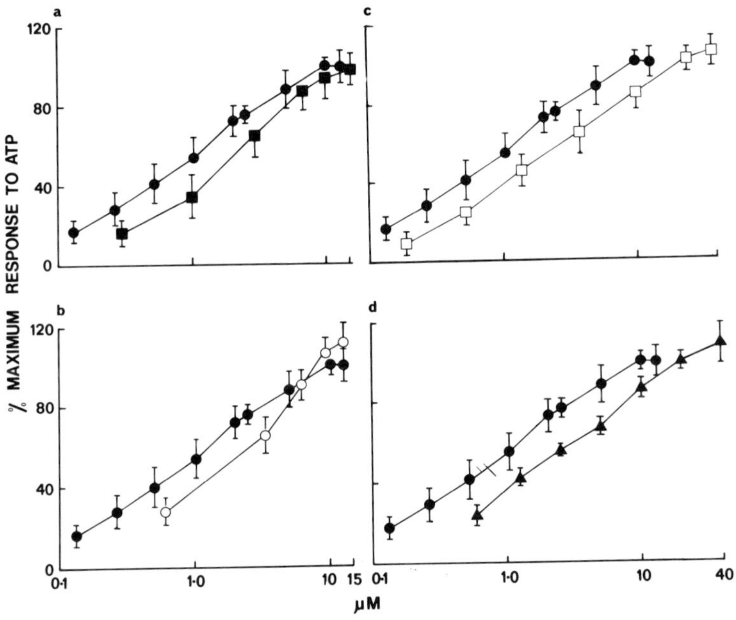

Fig 1. Relaxations of the guinea-pig taenia coli preparation, induced by the following: Panel (A); ATP (●) and 2'-deoxy-ATP (■); Panel (B); ATP (●) and 8-bromo-ATP (○); Panel (C); ATP (●) and 3'-deoxy-ATP (□); Panel (D); ATP (●) and ara-ATP (▲). Each point is the mean ± S.E. of at least six values obtained from at least two preparations and is expressed as a percentage of the maximum response to ATP.

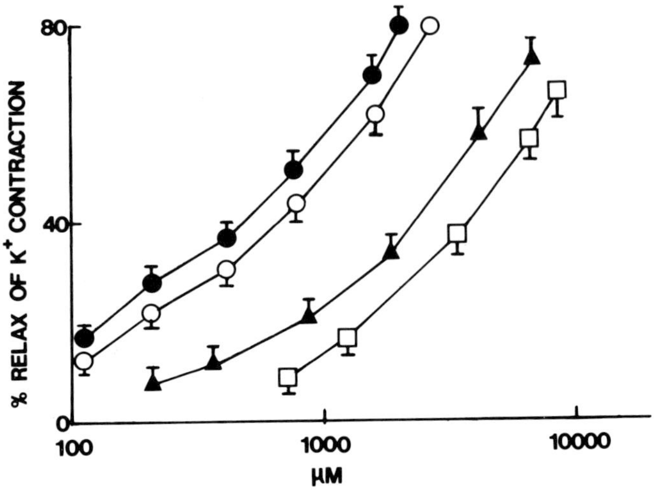

Fig 2. Relaxations of the rabbit portal vein preparation induced by the following: ATP (○); 8-bromo-ATP (●); ara-ATP (▲) and 3'-deoxy-ATP (□). 15 min prior to addition of drugs preparations were contracted in Krebs' solution with the potassium concentration adjusted to 167 mM. Each point is the mean ± S.E. of at least six values obtained from at least two preparations and is expressed as a percentage relaxation of the contraction due to potassium.

Thus in a number of different preparations, receptors for ATP and adenosine exhibit fundamental differences and in general ADP behaves like ATP and AMP like adenosine.

It cannot be stated with certainty that ATP has little or no action on P_1 receptors and that, similarly, adenosine has little or no action at P_2 sites. Reports to the contrary are quite frequent (see Maguire & Satchell 1981). However, such information must be interpreted with caution, expecially in relation to the actions of ATP at P_1 sites since ATP rapidly forms adenosine in many smooth muscle preparations.

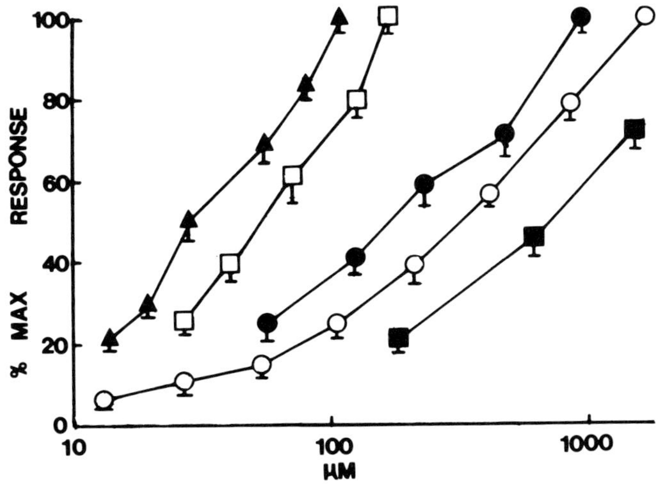

Fig 3. Contractions of the rat portal vein preparation induced by the following: ATP (○); 8-bromo-ATP (●); 2'-deoxy-ATP (■); 3'-deoxy-ATP (□) and ara-ATP (▲). Each point is the mean ± S.E. of at least six values obtained from at least two preparations and is expressed at a percentage of the maximum response to ATP.

Minimal Structural Requirements for Actions of P_1 and P_2 Receptors

The presence of a ribofuranose ring is essential for actions at P_1 receptors since adenine is usually inactive except in the guinea-pig trachea where it may be acting at a separate intracellular receptor (Satchell 1984). The 6-amino group is essential for actions of P_1 and P_2 since inosine, guanosine and their mono, di and tri phosphates have reduced potency on most smooth muscle containing preparations. Guanosine and its monophosphate was observed to be more potent than inosine and its monophosphate although less potent than adenosine and AMP in causing relaxation of guinea-pig taenia coli strips and potassium contracted rabbit portal vein (Satchell, unpublished observations) suggesting perhaps that the presence of the 2-amino group in guanosine could partially make up for the loss of the 6-amino group for action at P_1 sites.

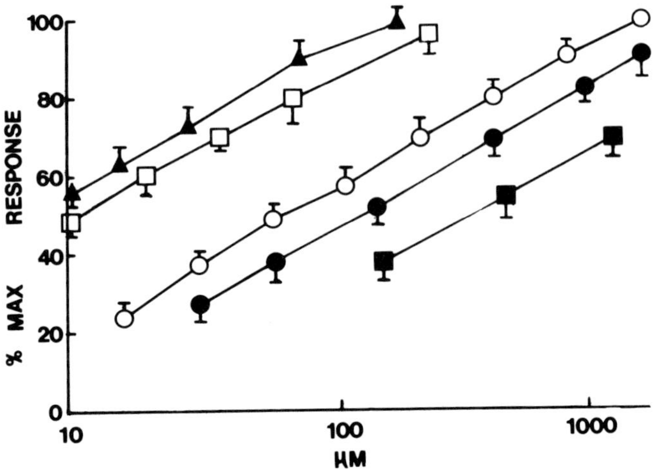

Fig 4. Contractions of rabbit femoral artery spiral strips induced by the following: ATP (○); 8-bromo-ATP (●); 2'-deoxy-ATP (■); 3'-deoxy-ATP (□) and ara-ATP (▲). Each point is the mean ± S.E. of at least seven values from at least three preparations and is expressed as a percentage of the maximum response to ATP.

The presence of a 5'-polyphosphate was mandatory for actions at P_2 sites, as discussed earlier there was not much difference in potency between two and three phosphates. Four phosphates were more active than three in causing relaxation of taenia coli strips (Satchell & Maguire 1975) and in causing contraction of the rat portal vein (Satchell, unpublished observations).

Discrete Properties of P_1 and P_2 Receptors

The degree of difference in the requirements for optimal agonist potency at P_1 and P_2 can be evaluated experimentally by making the same type of modification to the ATP and the adenosine molecule and examining agonist properties.

In guinea-pig taenia coli strips ara-adenosine and 8-bromoadenosine were inactive up to concentrations of 1 mM. In contrast to this, 8-bromo-ATP and ara-ATP were almost as potent as ATP (Fig. 1). 2' and 3'-deoxy-adenosine exhibited weak agonist properties with threshold relaxations at 300 µm (not shown) whereas 2' and 3'-deoxy-

ATP were almost as potent as ATP having similar dose-response curves (Fig. 1).

Earlier studies (Satchell & Maguire 1975) had shown that substitution at the 2 position of the purine moiety with a methythio group caused a 50-fold increase in inhibitory potency in taenia coli preparations. Whereas the same substitution in adenosine yielded a weak partial agonist.

Thus the divergent effects of identical structural modifications to ATP and adenosine emphasizes the point to which adenosine and ATP act exclusively at their receptors in this tissue.

<u>Discrete Properties of P_{2x} and P_{2y} Receptors</u>

Burnstock & Kennedy (1985) suggested that P_2 receptors could be separated into two subtypes on the basis of order of potency of ATP and its analogues and the actions of antagonists. P_{2x} receptors mediated excitation with a potency order: α,β-methylene ATP, β,γ-methylene ATP>ATP = 2-methythio ATP in the smooth muscle of vas deferens, bladder and arteries of several mammalian species. P_{2y} receptors mediated relaxation with a potency order: 2-methythio ATP >> ATP > α,β-methylene ATP, β,γ-methylene-ATP in guinea-pig taenia coli, rabbit portal vein and on vascular endothelial cells of several arteries, releasing endothelium-derived relaxing factor in the latter.

If P_{2x} and P_{2y} are discrete receptor populations with specific and different structural requirements for optimal agonist properties, then it was considered that this could be tested by examining the agonist properties of a series of ATP analogues on preparations containing the two receptor subtypes. Two homologous preparations were chosen, the longitudinal muscle of the rat portal vein for P_{2x} mediated contraction and the longitudinal muscle of the potassium-contracted rabbit portal vein for P_{2y} mediated relaxations.

In potassium contracted rabbit portal vein strips, 8-bromo-ATP was as potent as ATP, however ara-ATP and 3'-deoxy ATP were both substantially less potent than ATP (Fig. 2). In rat portal vein strips, 8-bromo-ATP was again as potent as ATP, but both ara-ATP and 3'-deoxy-ATP were almost an order of magnitude more potent than ATP (Fig. 3). Unlike 3'-deoxy-ATP, 2'-deoxy-ATP was less potent than ATP in the rat portal vein (Fig. 3).

Rabbit femoral artery spiral strips contract to ATP and its analogues suggesting the presence of P_{2x} receptors. This preparation is more sensitive to ATP than the rat portal vein and was used to further evaluate the relative potencies of the agonists. 8-bromo-ATP exhibited a similar but lesser potency than ATP in contrast to the rat portal vein study. Both ara-ATP and 3'-deoxy-ATP were again substantially more potent than ATP with $E.D_{50s}$ of approximately 10 μm (Fig. 4). Again 2'-deoxy-ATP was less potent than ATP in this preparation (Fig. 4).

The findings that ara-ATP and 3'-deoxy-ATP were both substantially less potent than ATP in causing relaxations in a preparation suggested to contain P_{2y}-receptors yet were both substantially more potent than ATP in two preparations suggested to contain P_{2x}-receptors supports the idea of 2 different P_2 receptor types with different structural requirements for agonist properties.

Concluding Remarks

Studies on the agonist properties of adenine nucleoside and nucleotide analogues in smooth muscle containing preparations demonstrate that the structural requirements for optimal agonist potency at P_1 and P_2 receptors are different and somewhat exclusive. Moreover, the studies support the view that P_2 receptors are heterogeneous and can validly be divided into two subclasses.

Acknowledgements

This work was carried out with the aid of a grant from the Australian Research Grants Scheme.

REFERENCES

Burnstock, G., 1978 A basis for distinguishing two types of purinergic receptor. In Cell Membrane Receptors for Drugs and Hormones: A Multidisiplinary Approach, edited by R. Straub & L. Bolis (Raven Press), p. 107.

Burnstock, G. & Brown, C., 1981 An introduction to purinergic receptors. In Purinergic Receptors, edited by G. Burnstock (Chapman and Hall Press), p. 1.

Burnstock, G. & Kennedy, C., 1985, General Pharmacology, 16, 433.

Holck, M. & Marks, B., 1978, Journal of Pharmacology and Experimental Therapeutics, 205, 104.

Maguire, M. & Satchell, D., 1981 Purinergic receptors in visceral smooth muscle. In Purinergic Receptors, edited by G. Burnstock (Chapman and Hall Press), p. 47.

Satchell, D., 1984), Trends in Pharmacological Science, 5, 340.

Satchell, D. & Maguire, M., 1975, Journal of Pharmacology and Experimental Therapeutics, 195, 540.

The role of ectonucleotidases in pharmacological responses to nucleotide analogues

N.J. CUSACK, S.M.O. HOURANI* & L.A. WELFORD

Department of Pharmacology, King's College London, Strand, London WC2R 2LS, U.K., and *Department of Biochemistry, University of Surrey, Guildford GU2 5XH, U.K.

INTRODUCTION

The pharmacological responses to adenine nucleotides in most tissues is transient, and this suggests that efficient mechanisms for the removal of ATP exist. It is generally accepted that ATP, being highly charged, does not cross cell membranes, and evidence from studies using ATP dual labelled in the adenine and in the phosphate groups suggests that ATP itself is not taken up intact, but is first broken down to adenosine, for which uptake systems exist (Burnstock 1972). The degradation of ATP to adenosine is now known to occur by sequential removal of phosphate groups to yield ADP, then AMP and finally adenosine, a process catalysed by a series of phosphohydrolases (ecto-nucleotidases) present on the outer surface of many cells (for reviews see Pearson 1986; Gordon 1986). The existence of a mechanism for the removal of ATP satisfies one of Eccle's criteria for a neurotransmitter, and so supports the suggestion that ATP may be an autonomic transmitter or co-transmitter (Burnstock 1986).

ATP causes contraction or relaxation of many smooth muscle preparations via P_2-purinoceptors, whereas adenosine nearly always causes relaxation, via P_1-purinoceptors, so that the breakdown of ATP to adenosine by ectonucleotidases may substantially modify the observed responses to ATP, which are therefore due to an undefined mixture of adenosine and adenine nucleotides. As analogues of ATP may be broken down by ectonucleotidases at different rates, observed structure-activity relationships may be influenced by such differences in the rates of dephosphorylation, and differences in pharmacological potencies may merely reflect differences in susceptibility to degradation. In this chapter we will discuss the ways in which modification of the structure of ATP can influence the rate of dephosphorylation of an analogue, and compare these structure-activity requirements for ectonucleotidases (figure 1) with the structure-activity requirements for pharmacological activity. Most of our structure-activity studies of ectonucleotidases (Welford et al. 1986; 1987) have been performed on two isolated smooth muscle preparations, the guinea-pig taenia coli in which ATP causes relaxation, and the guinea-pig urinary bladder in which ATP causes contraction. It has been suggested that these two effects of ATP may be mediated by two subtypes of the P_2-purinoceptor called P_{2Y} and P_{2X}, respectively (Burnstock & Kennedy 1985; for discussion see Cusack et al. this volume).

TAENIA COLI

Ectonucleotidases are present on isolated taenia coli muscle strips used in pharmacological studies, and dephosphorylation of exogenous ATP is rapid, being detectable after 1 minute and almost complete after 60 minutes, with adenosine as the major product (Hourani et al. 1985). At the highest concentration generally used for studies on their pharmacology (100 micromolar), ATP, ADP and AMP are all dephosphorylated at the same rate, and kinetic constants for the relevant ectoenzymes show ectoATPase to have a K_m of 230 micromolar and a V_{max} of 30 nanomoles per minute, the ectoADPase to have a K_m of 220 micromolar and a V_{max} of 20 nanomoles per minute, and the ectoAMPase (5'-nucleotidase) a K_m of 90 micromolar and a V_{max} of 3.4 nanomoles per minute (Welford et al. 1986). Removal of adenosine, presumably by uptake into the tissue, is slow with about 65% remaining after 60 minutes.

Sructure-activity requirements

Adenine-modified analogues of ATP. The other naturally occurring ribonucleoside triphosphates GTP, CTP and UTP are all dephosphorylated at the same rate as ATP showing that these ectonucleotidases are completely non-specific towards the base moiety, and tolerate pyrimidine as well as purine nucleotides. Substituents on the adenine ring of ATP has no effect on the rate of dephosphorylation, and N^6-phenyl-ATP, 8-bromo-ATP, 2-chloro-ATP and 2-methylthio-ATP are all degraded to the corresponding nucleosides as fast as ATP (Welford et al. 1986).

Ribose-modified analogues of adenine nucleotides. Replacement of the natural D-ribose sugar of adenine nucleotides by the unnatural L-ribose generates their L-enantiomers, and enables the stereoselectivity of the ectoenzymes to be determined (Holy & Sorm 1971). L-ATP is degraded

FIGURE 1. STRUCTURE-ACTIVITY REQUIREMENTS FOR RESISTANCE TO DEPHOSPHORYLATION BY ECTONUCLEOTIDASES

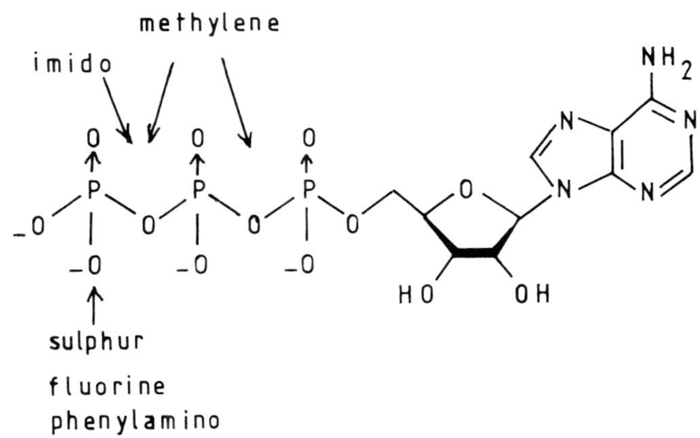

more slowly than ATP, about 30% remaining after 60 minutes, L-ADP is degraded very slowly with about 90% remaining after 60 minutes, and L-AMP is completely resistant. The 2-substituted enantiomers 2-chloro-L-ATP and 2-methylthio-L-ATP lost their terminal phosphates at the same rate as from L-ATP, but the composition of the breakdown products of the L-enantiomers reflects the stereoselectivity of the ectoADPase, so that the main products are L-ADP, 2-chloro-L-ADP and 2-methylthio-L-ADP respectively, rather than the adenosine analogues as in the case of the natural enantiomers (Welford et al. 1986).

Phosphate-modified analogues of adenine nucleotides. Replacement of an ionised oxygen by sulphur generates phosphorothioate analogues of adenine nucleotides. If the sulphur is attached to a non-terminal phosphate, then that phosphate is rendered chiral and the analogue exists as a pair of R_p and S_p diastereoisomers, which provide valuable information about the mechanism and stereoselectivity of dephosphorylation (Eckstein 1985). The ectoenzyme removes the terminal phosphate from ATP-α-S, an ATP analogue where the suphur is on the innermost phosphate, at the same rate as from ATP, and no stereoselectivity is exhibited. Removal of the terminal phosphate from the corresponding ADP analogue, ADP-α-S, is stereoselective and R_p-ADP-α-S is only slowly degraded, while the S_p diastereoisomer is degraded as rapidly as ADP to the AMP analogue, AMPS, which is resistant to further degradation to adenosine. Both isomers of ATP-β-S, where the sulphur is on the middle phosphate, lose their terminal phosphates more slowly than ATP, with the S_p diastereoisomer being more resistant than the R_p diastereoisomer, and the breakdown product in each case, ADP-β-S, is not degraded any further. ATP-γ-S, where the sulphur is on the terminal phosphate, is also not degraded (Welford et al. 1986).

Replacement of one of the bridging oxygens joined to phosphate with a methylene group generates a methylenephosphonate linkage which is completely resistant to chemical or enzymic hydrolysis (Myers et al. 1963). The methylene group can be inserted between the ribose sugar and the innermost phosphate as in homo-ATP (Jones & Moffett 1968), the innermost and middle phosphates as in α,β-methylene-ATP, and the middle and terminal phosphates as in β,γ-methylene-ATP. Homo-ATP loses its terminal phosphate at the same rate as ATP, but the corresponding ADP analogue, homo-ADP is only very slowly degraded to homo-AMP, which is immune to futher attack as expected. α,β-Methylene-ATP (AMPCPP) is slowly degraded to α,β-ADP, which is not degraded further. β,γ-Methylene-ATP (AMP-PCP), and its L-enantiomer β,γ-methylene-L-ATP (L-AMP-PCP) are completely resistant to dephosphorylation (Hourani et al. 1985; Welford et al. 1986).

Replacement of the electronegative oxygen linkage by the electropositive methylene linkage means that the resultant phosphonate analogues of ATP are not as fully ionised as ATP at physiological pH (Myers et al. 1963). To overcome this limitation, imido groups (Yount et al. 1971) and dihalomethylene groups (Blackburn et al. 1981) have been used instead of methylene to restore the electronegativity of the triphosphate chain. In spite of these improvements, β,γ-imido-ATP, β,γ-difluoromethylene-ATP and β,γ-dichloromethylene-ATP, like β,γ-methylene-ATP itself, are all completely resistant to degradation (Welford et al. 1986; Cusack et al. 1987). This indicates either the absence of ectopyrophosphatase or the inability of this enzyme to cleave imidodiphosphate or any of the methylenediphosphonates from these analogues.

Comparison of structure-activity requirements of ectonucleotidases with those of the inhibitory P_{2Y}-purinoceptor

Substitutions on the adenine ring have virtually no effect on the rate of removal of the terminal phosphate from the ATP analgues, but they have widely differing potencies as relaxants of the guinea-pig taenia coli with activities relative to ATP ranging from 200 for 2-methylthio-ATP, 35 for 2-chloro-ATP and 0.4 for 8-bromo-ATP (Welford et al. 1986; Burnstock et al. 1983; Satchell & Maguire 1982) while N^6-phenyl-ATP is inactive (Cusack & Hourani unpublished). Little stereoselectivity exists for the dephosphorylation of the enantiomers of ATP and this is unchanged by 2-substitution (Welford et al. 1986), but 2-substituents have a profound effect on the pharmacological responses (Cusack & Planker 1979), with 2-chloro-ATP and 2-methylthio-ATP being 54 times and over 700 times, respectively, as potent as their corresponding L-enantiomers (Burnstock et al. 1983). Dephosphorylation of the phosphorothioate analogues of ATP is stereoselective only if the sulphur is attached to the middle phosphate as in ATP-β-S, whereas stereoselectivity for induction of relaxation is seen only when the suphur is attached to the innermost phosphate as in ATP-α-S. ATP-β-S is degraded more slowly than ATP but is no more potent than ATP at causing relaxation, while ATP-α-S is rapidly degraded but is much more potent than ATP at relaxing the taenia coli. Again, ATP-γ-S is completely resistant to degradation but is only as potent as ATP (Welford et al. 1986; Burnstock et al.

1984). The effect of replacing an oxygen group by methylene depends on the point of insertion into the triphosphate chain. Homo-ATP loses its terminal phosphate as fast as ATP and is 70 times as potent as ATP at relaxing the taenia coli, α,β-methylene-ATP (AMPCPP) is only slowly degraded and is approximately equipotent with ATP as a relaxant, while β,γ-methylene-ATP (AMP-PCP) is not degraded and is only 0.2 times as potent as ATP, and its L-enantiomer β,γ-methylene-L-ATP (L-AMP-PCP) is neither degraded nor active on the taenia coli (Welford et al. 1986; Maguire & Satchell 1979; Hourani et al. 1985).

Overall, these studies show that the structure-activity requirements of ectonucleotidases on the isolated guinea-pig taenia coli do not resemble those of the inhibitory P_{2Y}-purinoceptor, and suggest that neither dephosphorylation nor resistance to dephosphorylation is a prerequisite for induction by adenine nucleotides of relaxation of the taenia coli (Welford et al. 1986).

URINARY BLADDER

Ectonucleotidases are also present on the isolated urinary bladder muscle strips used in pharmacological studies, and dephosphorylation of exogenous ATP to adenosine is rapid being detectable after 1 minute and complete after 60 minutes (Cusack & Hourani 1984). Kinetic studies show the ectoATPase to have a K_m of 532 micromolar and a V_{max} of 1.1 nmol. per minute, the ectoADPase to have a K_m of 864 micromolar and a V_{max} of 0.14 nanomoles per minute, but the ectoAMPase is not saturable even at an AMP concentration of over 6 millimolar (Welford et al. 1987).

Comparison of structure-activity requirements of ectonucleotidases with those of the excitatory P_{2X}-purinoceptor

Substitutions on the adenine ring have virtually no effect on the rate of removal of the terminal phosphate of the ATP analogues, nor do they have any effect on the potencies of the analogues at inducing contraction of the urinary bladder. Little stereoselectivity exists for the dephosphorylation of the enantiomers of ATP and this is unchanged by 2-substitution, and little or no effect is seen on the pharmacological response (Welford et al. 1987; Burnstock et al. 1983). Dephosphorylation of the phosphorothioate analogues is stereoselective only towards the diastereoisomers of ATP-β-S, and stereoselectivity for contraction of the bladder is exibited towards the more resistant S_p isomer of ATP-β-S. Both isomers of ATP-β-S are degraded more slowly than ATP, and both are much more potent than ATP at contracting the bladder, while ATP-α-S is degraded as rapidly as ATP and is no more potent than ATP at contracting the urinary bladder. ATP-γ-S is only slowly degraded and is considerably more potent than ATP at contracting the bladder (Welford at al. 1987; Burnstock et al. 1984) The effect of replacing a bridging oxygen in the triphosphate chain by methylene depends on the point of insertion. Homo-ATP loses its terminal phosphate as rapidly as ATP, and is only as potent as ATP at contracting the bladder, α,β-methylene-ATP is very slowly degraded and is much more potent than ATP, while β,γ-methylene-ATP and β,γ-imido-ATP are completely resistant to degradation and have high pharmacological potency (Welford et al. 1987; Lucacsko & Krell 1982). β,γ-methylene-L-ATP (L-AMP-PCP) and its isopolar analogue β,γ-difluoromethylene-L-ATP are both completely resistant to dephosphorylation, and these analogues have the highest known potencies at the excitatory P_{2X}-purinoceptor (Cusack & Hourani 1984; Cusack et al. 1987).

Our results show that there is a strong correlation between the structure-activity requirements of ecto-nucleotidases on the guinea-pig isolated urinary bladder and those of the excitatory P_{2X}-purinoceptor, that ATP analogues that are rapidly degraded have only limited pharmacological activity, and that in general analogues of ATP that are resistant to dephosphorylation have a greatly enhanced ability to induce contraction of the guinea-pig urinary bladder (Welford et al. 1987).

CONCLUSIONS

The results from these studies on the structure-activity requirements of ectoenzymes present on the taenia coli and on the urinary bladder show that they are very similar, and they are also similar to those of ectonucleotidases on endothelial and vascular smooth muscle cells in culture (see Gordon 1986). The importance of ectoenzymes in neurogenic responses has recently been heightened by the analysis of their role in providing adenosine from adenine nucleotides released following nerve stimulation of immunoisolated cholinergic synapses (Richardson et al. 1987) and in the rat vas deferens (Webster et al. 1986).

REFERENCES

Blackburn, G.M., Kent, D.E. & Kolkmann, F., 1981, Journal of the Chemical Society chemical Communications, 1188.
Burnstock, 1986, Acta Physiologica Scandinavia, 126, 67.
Burnstock, G., 1972, Pharmacological Reviews, 24, 509.
Burnstock, G., Cusack, N.J., Hills, J.M., Mackenzie, I. &

Meghji, P., 1983, British Journal of Pharmacology, 79, 907.

Burnstock, G. & Kennedy, C., 1985, General Pharmacology, 16, 433.

Burnstock, G., Cusack, N.J. & Meldrum, L.A., 1984, British Journal of Pharmacology, 82, 369.

Cusack, N.J. & Hourani, S.M.O., 1984, British Journal of Pharmacology, 82, 155.

Cusack, N.J., Hourani, S.M.O., Loizou, G.D. & Welford, L.A., 1987, British Journal of Pharmacology, 90, 791.

Cusack, N.J. & Planker, M., 1979, British Journal of Pharmacology, 67, 153.

Eckstein, F., 1985, Annual Reviews of Biochemistry, 54, 367.

Gordon, J.L., 1986, Biochemistry Journal, 233, 309.

Holy, A. & Sorm, F., 1971, Collected Czechoslovac Chemical Communications English Edition, 36, 3282.

Hourani, S.M.O., Cusack, N.J. & Welford, L.A., 1985, European Journal of Pharmacology, 108, 197.

Jones, G.H. & Moffatt, J.G., 1963, Journal of the American Chemical Society, 90, 5337.

Lucacsko, P. & Krell, R.D., 1982, European Journal of Pharmacology, 80, 401.

Maguire, M.H. & Satchell, D.G., 1979, specificity of adenine nucleotide receptor sites: inhibition of guinea-pig taenia coli by adenine nucleotide analogues. In Physiological and Regulatory Functions of Adenosine and Adenine Nucleotides, edited by H.P. Baer and G.I. Drummond (Raven Press) p. 33.

Myers, T.C., Nakamura, K. & Fleshner, J.W., 1963, Journal of the American Chemical Society, 85, 3292.

Pearson, J.D., 1986, ectonucleotidases of vascular endothelial cells: characterisation and possible physiological roles. In Cellular Biology of Ectoenzymes, edited by G.W. Kreutzberg (Springer-Verlag) p. 17.

Richardson, P.J., Brown, S.J., Bailyes, E.M. & Luzio, J.P., 1987, Nature, 327, 232.

Satchell, D.G. & Maguire, M.H., 1982, European Journal of pharmacology, 81, 669.

Webster, D.R., Boston, G.D., Holford, H.G. & Paton, D.M., 1986, Naunyn-Schiedeberg's Archives of Pharmacology, 333, 163.

Welford, L.A., Cusack, N.J. & Hourani, S.M.O., 1986, European Journal of Pharmacology, 129, 217.

Welford, L.A., Cusack, N.J. & Hourani, S.M.O., 1987, European Journal of Pharmacology, in the press.

Yount, R.G., Babcock, D., Ballantyne, W. & Ojali, D., 1971, Biochemistry, 10, 2484.

MECHANISMS OF ACTION OF ADENOSINE

PHYSIOLOGICAL MODULATION OF SYNAPTIC TRANSMISSION AND NEURONAL CALCIUM INFLUX BY ADENOSINE

Peter Schubert

Max Planck Institute for Psychiatry, Department of Neuromorphology, am Klopferspitz 18a, 8033 Martinsried, Fed. Rep. Germany

Modulation of nerve cell function by adenosine seems to be the complex result of a concerted action of different effects. They all are receptor -mediated and the A1-receptor seems to be of particular importance (Reddington et al. 1982; Dunwiddie and Fredholm 1984). In an attempt to localize by EM -autoradiography the binding sites of an A1 agonist we found that they were not only associated with synaptic complexes but also with extrasynaptic sites (Tetzlaff et al. 1987). Accordingly, synaptic and non -synaptic adenosine actions can be distinguished. At the synapse, adenosine depresses the signal transmission in a concentration -dependent manner thus modulating the amount of afferent activation addressed to a neuron. In addition, adenosine may also influence the evoked neuronal firing pattern elicited in response to afferent activation; it antagonizes the generation of repetitive discharges. This 'antiepileptic' effect is seen in the absence of synaptic transmission and reflects a non -synaptic modulation of the soma membrane properties (Schubert and Lee 1986).

Most of these adenosine effects reported were observed at concentrations of 20 -40 uM which may be reached in the extracellular space only under pathological conditions as, for example, during brain ischaemia. However, the endogenous extracellular adenosine concentration in the normally functioning brain is about 1 µM (Zetterstrom et al. 1982) and it is difficult to extrapolate from high dosage effects on the modulation exerted physiologically by adenosine in the low concentration range.

This is subject of the present paper and special emphasis is paid to the modulation of synaptic transmission and stimulus evoked neuronal calcium influx. Both are presumably of major importance for determining the signal processing in a neuronal network. In this respect, the phenomenon of synaptic plasticity has attracted particular interest and the regulatory mechanisms underlying these long lasting changes of synaptic transmission. Several mechanisms seem to be involved in the regulation of the efficiency of synaptic transmission. There is growing evidence that an activation of the protein kinase C (PKC) is of central importance for the tuning of synaptic function (see e.g. Malenka et al. 1986, 1987). In order to test whether the different control mechanisms might be interconnected and adjusted to each other, we studied wether an activation of PKC, as can be achieved in the experiment by phorbol esters, goes along with a change of the modulatory action exerted by adenosine.

The experiments were performed in rat hippocampal slices (fig.1). Changes in the efficiency of synaptic transmission were determined by measuring the calcium fluxes elicited in the synaptic region and in the soma layer of CA1 pyramidal neurons upon activation of the afferent fibers by a 20 Hz/10 sec stimulus train. The decrease in the extracellular

Ca^{2+}concentration measured in the soma layer is thought to represent the action potential induced Ca^{2+} entry into the pyramidal cell bodies through voltage -dependent ion channels. This was

Figure 1. Technical procedure

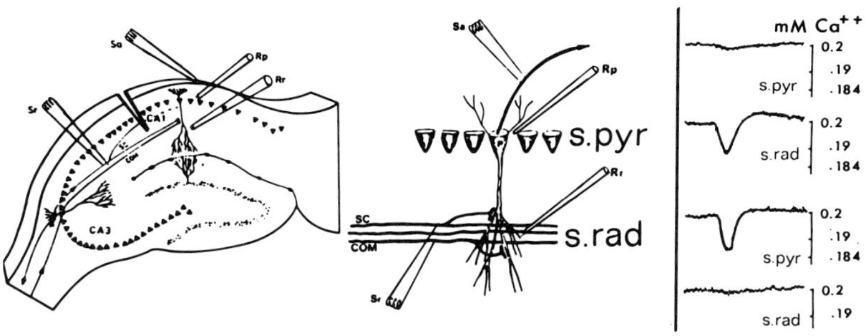

Rat hippocampal slices superfused with a medium containing (in mM): $CaCl_2$, as indicated, $MgSO_4$ 2.4, KCl 3.3, NaCl 124, $NaHCO_3$ 25.7, KH_2PO_4 1.25, glucose 10. Stimulus -evoked decreases in the extracellular Ca^{2+}concentration (Ca^{2+} signals) were measured with two ion sensitive/recording electrodes (for preparation see Heinemann et al. 1977), located in S. rad and S. pyr. above traces: In 0.2 mM Ca^{2+} medium when synaptic transmission is blocked, orthodromic stimulation of the afferent radiatum fibers by a 20 Hz/10 sec stimulus train leads to a marked Ca^{2+}signal in the synaptic region (S.rad), but not in the soma layer of the postsynaptic pyramidal neurons (s.pyr). below traces: Direct antidromic stimulation of the pyramidal neurons generates a Ca^{2+} signal measured in S pyr reflecting somatic Ca^{2+} influx which is not seen and does not interfer with the measurements of the electrode in S.rad.

found to be a rather sensitive measure for the amount of activation elicited in CA1 neurons upon synaptic activation. Thus, during recovery from a blockade of synaptic transmission large Ca^{2+} signals appeared in the soma layer already at a time point when the number of activated postsynaptic neurons was still too small or the evoked activity not sufficiently synchronized to be detected as extracellularly recordable field potentials (Schubert and Kreutzberg 1987).

The physiological action of adenosine was tested by evaluating the effect of removing a possible action by the extracellularly present concentration of 1 μM. This was done in two ways: 1. by antagonizing receptor -binding with the antagonist theophylline and 2. by enzymatic degradation of the endogeneous adenosine with deaminase (ADA) added to the superfused medium. In order to titrate the strenghtening of synaptic transmission resulting from the removal of a tonic adenosine action, some of the experiments were performed at a reduced critical Ca^{2+} level of 0.4 mM at which an almost complete labile blockade of synaptic transmission is achieved. This is known from reported intracellular recordings (e.g. Jones and Heinemann 1987) and also indicated by the absence of a significant decrease of extracellular Ca^{2+} measured in the pyramidal soma layer upon synaptic activation (fig. 2).

Upon removal of a possible adenosine action by ADA or theophylline, a significant stimulus -evoked calcium signal appeared in S.pyr. Field potentials were never seen upon single pulses but were sometimes generated late during the stimulus train - a sign of frequency potentiation pointing to a presynaptic site of action. Such a recovery of synaptic transmission and evoked postsynaptic Ca^{2+} influx could be even achieved if the

Ca^{2+} concentration in the medium was further reduced to 0.2 mM , but only if the removal of adenosine action occured in conjunction with a K^+ channel blocker allowing by broadening the action potential duration an enhanced presynaptic Ca^{2+} influx (Schubert and Heinemann 1987). These findings demonstrate a powerful control of synaptic transmission by physiological adenosine concentrations and a plausible explanation for this apparently calcium -dependent effect is that adenosine acts by decreasing the sensitivity of transmitter release to calcium.

PKC is thought to act on the same mechanism. There is evidence from non neuronal systems that phorbol esters increase the calcium sensitivity of secretory processes (Baker 1984) and this may also explain the observed enhancement of transmitter release and of evoked potentials in the hippocampus (Malenka et al. 1986, 1987).

Figure 2. Effect of theophylline (theo, 50 µM) on the synaptic and somatic Ca^{2+} influx evoked upon afferent stimulation (left); antagonism by phorbol esters (right).

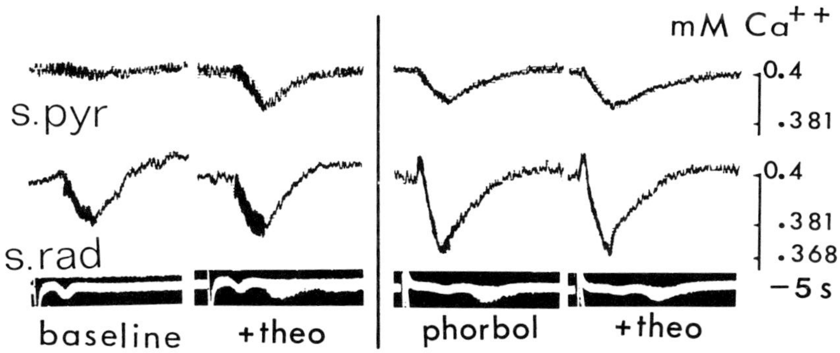

Accordingly, pretreatment of hippocampal slices with 1 μM phorbol-12,13-dibutyrate increases markedly the stimulus -evoked neuronal Ca^{2+} influx both, in normal as well as in low Ca^{2+} medium (figs. 2 and 3). Similar to the effect of removing physiological adenosine action, a recovery of synaptic transmission is also achieved by phorbol esters in 0.4 mM Ca^{2+} medium as indicated by the stimulus -evoked somatic Ca^{2+} influx and by the recording of field potentials. However, theophylline is no more effective under these conditions and the stimulus -evoked Ca^{2+} signals remain unchanged (Schubert 1987a; fig. 2). Thus, the possibility of a physiological modulation of synaptic transmission by the endogenous extracellular adenosine concentration seems to be blocked upon activation of PKC. A similar effect, i.e. a reduction of the theophylline effectivity, was observed when long term potentiation had been generated in a pathway by high frequency potentiation suggesting that similar mechanisms are involved (Schubert 1987b).

The depressory effect of enlarged adenosine concentrations of 40 μM on synaptic transmission and neuronal Ca^{2+} influx was also blocked by phorbol esters if tested in normal (2mM) Ca^{2+} medium (fig. 3). In this context it is interesting to note that also another A1 receptor -mediated adenosine effect, i.e. the inhibition of cAMP synthesis has been recently reported to be blocked by phorbol esters (Norsted and Fredholm 1987).

Also the stimulus -evoked dendritic Ca^{2+} influx which is usually even more increased by phorbol esters than the somatic Ca^{2+} signal shows no response to adenosine. Since the enlarged dendritic Ca^{2+} signal includes an APV (2-amino-5-phosphono-valeric acid) -sensitive component,

Figure 3. Antagonism of 'high dose' adenosine action in 2.0 mM Ca^{2+} medium by phorbol esters.

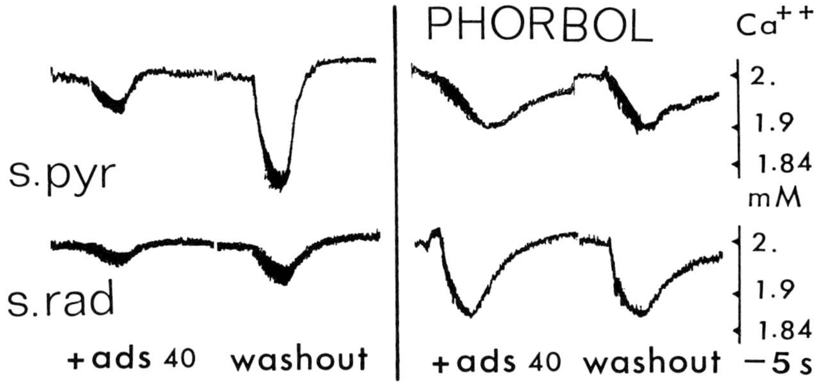

Figure 4. Synaptic Ca^{2+} influx elicited by orthdromic stimulation in 0 Mg^{2+}/normal Ca^{2+} -medium' is increased by theophylline (left); the effect is absent in the presence of 50 μM APV removing the NMDA receptor -mediated component of Ca^{2+} influx.

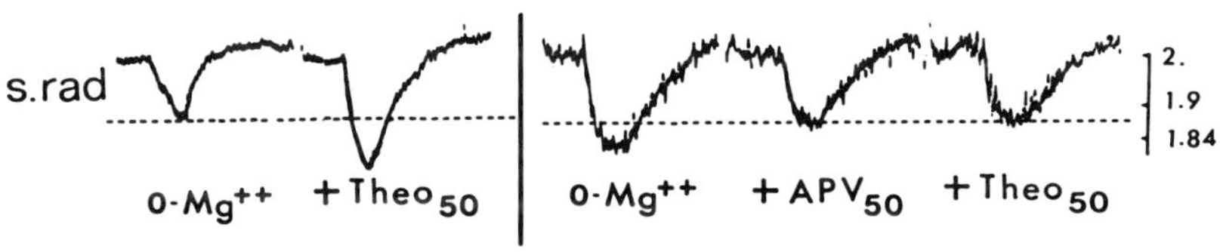

the finding means that also this NMDA-receptor -mediated neuronal Ca^{2+} influx is not affected by adenosine. On the other hand, if tested in the absence of phorbol esters under conditions when NMDA-receptors are activated, these receptor -mediated synaptic Ca^{2+} fluxes seem to some extent controlled by physiological adenosine concentrations (fig. 4; Schubert 1987c).

REFERENCES

Baker, P. F., 1984, Nature, 308, 693.

Dunwiddie, T. and Fredholm, B., 1984, Naunyn Schmiedebergs Archiv of Pharmacology, 326, 249.

Heinemann, U., Lux, H.D. and Gutnick, M.J., 1977, Experimental Brain Research, 27, 237.

Jones, R. S., G. and Heinemann, U., 1987, Experimental Brain Research, 65, 676.

Malenka, R.C., Madison, D. and Nicoll, R.A., 1986, Nature, 321,157.

Malenka, R.C., Ayoub, G.S. and Nicoll, R.A., 1987, Brain Research, 403, 198.

Norstedt, C. and Fredholm, B.F., 1987, Naunyn Schmiedeberg's Archiv of Pharmacology, 335, 136.

Reddington, M., Lee, K.S. and Schubert, P., 1982, Neuroscience Letters, 28,275.

Schubert, P., 1987a, submitted.

Schubert, P., 1987b, Proceedings of the IBRO Symposium on Synaptic Plasticity in the hippocampus, p. 19.

Schubert, P., 1987c, submitted.

Schubert, P. and Heinemann, U., 1987, submitted.

Schubert, P. and Kreutzberg,G.W., 1987, In Topics in Adenosine Research, edited by E. Gerlach and B. Becker, (Springer Verlag), p. 521.

Schubert, P. and Lee, K., 1986, Neuroscience Letters, 67, 334.

Tetzlaff, W., Schubert, P. and Kreutzberg, G. W., 1987, Neuroscience, 27, 869.

Zetterstrom, T., Vernet, L., Ungerstedt, U., Jonzon, B., Tossmann, U. and Fredholm, B.B., 1982, Neuroscience Letters, 29, 111.

Acknowledgement

I thank Prof. G.W. Kreutzberg for his continuous encouragement of these studies and Maria Koeber for skillful technical assistence.

MECHANISMS UNDERLYING ADENOSINE MODULATION OF HIPPOCAMPAL EXCITATORY TRANSMISSION

Thomas V. Dunwiddie and William R. Proctor

University of Colorado Health Sciences Center
4200 E. 9th Ave., Denver, Colorado 80262
and
VA Medical Research Service,1055 Clermont St.
Denver, CO 80220

Adenosine has a number of well-characterized modulatory actions upon the physiological activity of the central nervous system, and appears to act at multiple sites to elicit these responses. From a physiological standpoint, adenosine has been shown to modulate synaptic transmission at excitatory and inhibitory synapses by reducing transmitter release. Although this latter phenomenon was originally noted in the peripheral nervous system, the ability to depress synaptic transmission has now been described in the central nervous system in the olfactory cortex (Okada and Kuroda, 1975, 1980; Scholfield, 1978) cerebellum (Kocsis et al, 1984), and hippocampus (Schubert and Mitzdorf, 1979; Dunwiddie and Hoffer, 1980; Okada and Ozawa, 1980). What is still unclear is whether the same mechanism(s) that underlie the postsynaptic effects of adenosine, such as activation of potassium conductance, and inhibition of calcium currents, contribute to the modulation of transmission, or whether adenosine receptors located upon presynaptic terminals act via entirely different mechanisms to modulate synaptic efficacy.

As in the peripheral nervous system, the inhibitory effect of adenosine reflects a reduction in transmitter release from the nerve terminal. There are three likely hypotheses as to the mechanism by which adenosine modulates synaptic transmission. The first suggests that adenosine acts upon extracellular receptors, which in turn act via a second messenger system to directly affect the mechanism by which transmitter is released. An alternative hypothesis is that adenosine inhibits the influx of calcium via the voltage sensitive calcium channels that are located on presynaptic nerve terminals. A third hypothesis is that adenosine

Abbreviations used: 4 AP, 4-aminopyridine; LCB, low calcium bursting; TEA, tetraethylammonium; NE, norepinephrine; PIA, phenylisopropyl adenosine; AMP, adenosine monophosphate

can open potassium channels in the presynaptic terminal, and that this consequently reduces the influx of calcium, and hence transmitter release.

The most compelling evidence in support of the first hypothesis comes from studies of adenosine actions at the neuromuscular junction (see Silinsky, 1986, Silinsky et al, 1987 for reviews). These studies have shown that the efflux of transmitter that occurs under conditions when calcium influx into the nerve terminal through voltage sensitive channels is precluded, is still sensitive to the inhibitory effects of adenosine. Although there is good evidence to support this hypothesis, it is difficult to determine whether a similar mechanism exists in the central nervous system. Because there are few central synapses at which measurements of mepps and quantal release can be made, the generality of this mechanism remains unclear.

With regards to the other hypothetical mechanisms of synaptic modulation, it has been quite difficult to provide direct evidence in support of these hypotheses, and this can be largely attributed to the difficulties in making direct measurements of ionic fluxes in presynaptic nerve terminals. The primary experimental strategies that have been employed have been to characterize adenosine responses in cells from which direct measurements <u>can</u> be made (e.g., intracellular recording from cell bodies), and then to compare these responses to presynaptic effects of adenosine. More specifically, such responses can be compared on the basis of a) the pharmacological properties of the receptors involved, b) the biochemical mechanisms (if any) that are linked to these receptors, and c) the ion channels involved, and their sensitivity to agents that can block specific post-synaptic ion channels. Each of these approaches can provide only indirect evidence supporting a role for specific cellular mechanisms; however, in the remainder of this chapter, we would like to evaluate the evidence to support these putative mechanisms of action.

For the sake of such comparisons, we have focused on three such postsynaptic actions of adenosine, each of which could reduce transmitter release should an analogous process be found in the presynaptic terminal. These actions are 1) the direct inhibition of a calcium current that has been described in dorsal root ganglion cells (Dolphin et al, 1986; Macdonald et al, 1986), in sympathetic neurons (Henon and McAfee, 1983), and in hippocampal CA3 neurons (Madison, Fox and Tsien, 1987); 2) the hyperpolarization of postsynaptic neurons by the activation of a potassium current (e.g., Segal, 1982; Trussell and Jackson, 1985), and 3) the inhibition of low calcium-induced bursting (LCB) in the rat hippocampus following antidromic stimulation (Lee et al, 1984; Haas et al, 1984). Furthermore, comparisons will be limited primarily to the modulation of excitatory amino acidergic transmission in the CA1 region of the hippocampus and in the olfactory cortex. The comparison of pre- and postsynaptic actions of adenosine is outlined in Table I.

One criterion upon which to identify a putative mechanism of presynaptic adenosine action is the pharmacological properties of the modulatory receptors. At most synapses, an adenosine receptor similar to the A1 subtype is involved (Dunwiddie, 1985), and this appears to be the case in hippocampus and olfactory cortex as well (Reddington et al, 1982; McCabe and Scholfield, 1985; Dunwiddie and Fredholm, 1984; Dunwiddie et al, 1986). However, although the pharmacology is consistent with receptors of the A1 subtype, direct tests for an involvement of adenylate cyclase have been largely inconclusive. Cyclic AMP analogs do not reverse the inhibition of transmission mediated by adenosine in hippocampus (Dunwiddie and Fredholm, 1985). Forskolin, which activates adenylate cyclase, can partially reverse adenosine mediated depression of transmission in olfactory cortex (McCabe and Scholfield, 1985), but cannot reverse the inhibition of NE release in hippocampal slices by PIA (Fredholm and Lindgren, 1987). Therefore, it would be difficult to support the hypothesis that adenosine acts <u>only</u> via inhibition of cyclase to inhibit release. A point that should be stressed is that while cyclic AMP may well affect transmitter release, this does not necessarily imply that adenosine must also modulate release through such a mechanism, particularly in view of the recent observations that adenosine can elicit physiological responses without any involvement of adenylate cyclase or cyclic nucleotides (see below).

In terms of the receptors that mediate the three postsynaptic effects of adenosine (see Table I), a lack of pharmacological data makes comparisons difficult. The inhibition of low calcium-induced bursting (LCB) by adenosine appears to involve an A1-like receptor, based upon the relatively high potency of R-PIA vs. D-PIA (Lee et al, 1984), and the similar effects of the A1-selective selective antagonist cyclopentyltheophylline in the two systems (Dunwiddie, unpublished). There have been no quantitative pharmacological studies that permit the identification of the receptor that is linked to potassium channels. There is also relatively little information concerning the receptor that inhibits calcium currents; in dorsal root ganglion cells, PIA, CHA, and 2CA were found to be relatively potent, suggesting an A1 receptor, but NECA was completely inactive (Macdonald et al, 1986), which is inconsistent with the actions of NECA on neurotransmission, and with its affinities for either the A1 or A2 receptor.

Because the postsynaptic hyperpolarization and decrease in cellular input resistance reflect the activation of a potassium conductance, another basis for comparing pre- and postsynaptic responses is upon sensitivity to antagonism by potassium channel blockers. The hyperpolarizing response to adenosine in

	Postsynaptic hyperpolarizations	Low calcium induced bursting	Presynaptic modulation of transmission	Calcium Currents
Pharmacology	Not well characterized	A1[11,16]	A1 in hippocampus[6,20] A1-like in olfactory cortex, but NECA>PIA=CHA[21]	Similar to A1[17], except NECA ineffective
GTP sensitive	Yes[27]	Not known	Not known	Not known in brain, yes in neuroblastoma X glioma cells[15]
NEM sensitive	Not known	Not known	Yes[7];also true for $alpha_2$ and adenosine modulation of NE release[1,12]	Not known
4AP sensitive	No[13,24]	Yes[23]	Yes[9,21,22,26]	Not known
TEA sensitive	No[13]	No[23]	No[22]	No[5,17]
Cesium sensitive	Not known?	Not known	Yes[22] (extracellular cesium)	No[5]
Barium sensitive	Yes[10]	Not known	No[21,22]	No[5,14]
Pertussis toxin	Yes[2,10,27]	Yes[11]	No[11]; Yes in cultured cerebellar neurons[4]	Not known
cAMP	Not involved[27]		Yes at NMJ and cerebellum[3,25] No effect of 8mtcAMP in hippocampus[8] Forskolin partially reverses adenosine in olfactory cortex[21] Forskolin affects basal NE release, but not PIA inhib.[12]	Unclear; effect of 8brcAMP blocked by IBMX[5]

1 Allgaier et al, 1987
2 Bohm et al, 1986
3 Dolphin and Archer, 1983
4 Dolphin and Prestwich, 1985
5 Dolphin et al, 1986
6 Dunwiddie and Fredholm, 1984
7 Dunwiddie and Fredholm, unpublished
8 Dunwiddie and Fredholm, 1985
9 Dunwiddie and Proctor, 1987
10 Dunwiddie and Proctor, unpublished
11 Dunwiddie, unpublished
12 Fredholm and Lindgren, 1986
13 Greene and Haas, 1985
14 Haas and Greene, 1984
15 Hescheler et al, 1987
16 Lee et al, 1984
17 Macdonald et al, 1986
18 Madison et al, 1987
19 McCabe and Scholfield, 1985
20 Reddington et al, 1982
21 Scholfield and Steel, 1987
22 Scholfield, 1986
23 Schubert and Lee, 1986
24 Segal, 1982
25 Silinsky, 1984
26 Stone, 1981
27 Trussell and Jackson, 1986

hippocampal and striatal neurons results from the activation of a potassium channel that is coupled to an adenosine receptor via a GTP-binding intermediate (G_i, G_0, or an unknown G protein), and this potassium channel is insensitive to both TEA and 4AP (Segal, 1982; Greene and Haas, 1985), but can be blocked by barium (0.5 mM; fig. 1). In contrast, the effect of adenosine on LCB is completely antagonized by low concentrations (50-100 uM) of 4AP (Schubert and Lee, 1986). The effects of adenosine upon calcium currents and spikes are unaffected by TEA, cesium, or barium (Haas and Greene, 1984; Macdonald et al, 1986; Dolphin et al, 1986). The presynaptic effects of adenosine appear to be most similar to the blockade of LCB, in that they are markedly attenuated by aminopyridines (Stone, 1981; Dunwiddie and Proctor, 1987;

Fig. 1 **Effects of potassium channel blockers on adenosine responses.** The effect of intracellular cesium (2M CsCl in the recording pipette) or extracellular barium (0.5 mM) on adenosine responses is illustrated. For each measure, responses obtained in control medium with KCl or K acetate electrodes were normalized to 100%, and the absolute value of the control responses (in mV or percent change from pre-adenosine control) is noted below the group of columns. The response measures were, from left to right; the hyperpolarization induced in the resting membrane potential, the decrease in the cellular input resistance, and the percent decrease in the intracellularly recorded epsp, all in response to 50 uM adenosine. The only significant effect in each case was elicited by barium ($p<.05$ for each response). The lack of any effect of cesium may reflect a failure to diffuse out to dendritic sites of adenosine action.

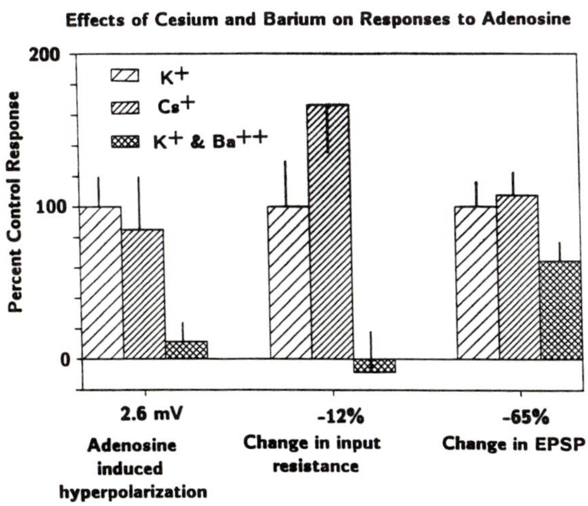

Scholfield and Steel, 1987), but not by TEA. It has also been reported that extracellular cesium can also block the presynaptic effects of adenosine, but that such responses are insensitive to barium (Scholfield and Steel, 1987).

A potential problem lies in the interpretation of experiments involving potassium channel blockers; agents such as 4AP markedly increase the presynaptic fiber spike, and by themselves potentiate synaptic transmission, most probably by increasing the influx of calcium into the presynaptic nerve terminal. Under such conditions, it is possible that the sensitivity to adenosine is lost because the additional influx of calcium saturates the release mechanism. If adenosine acts by reducing calcium influx, or diminishes the sensitivity of the release process to calcium, such a mechanism might be blocked by high concentrations of intraterminal calcium. Scholfield and Steel (1987) controlled for this possibility by altering the divalent cation concentrations in their medium so that the response amplitude in 4AP is the same as that prior to 4AP treatment. Under these conditions, the effect of adenosine is still almost entirely blocked by 4AP. Nevertheless, it is difficult to conclude unequivocally from these experiments that the primary effect of adenosine is upon a 4AP sensitive potassium channel. Because 4AP prolongs the presynaptic action potential, it is at possible that transmitter release may shift in its dependence upon one calcium current (most likely the N current) that is normally responsible for the calcium influx that leads to transmitter release (Perney et al, 1986; Miller 1987), to another current that may be insensitive to adenosine (e.g., the L current).

A final basis upon which comparisons between different adenosine responses can be drawn is their apparent dependence upon G proteins as intermediaries in the physiological response. One way to test for such a mechanism is to determine whether intracellular GTP is required in order to observe the response. The only situation in which this has been shown to be the case is for the activation of the potassium conductance by adenosine in striatal and hippocampal neurons in culture; Trussell and Jackson (1985, 1986) demonstrated that when such cells are recorded with the whole cell clamp technique, responses to adenosine are gradually lost unless GTP in included in the recording pipette.

A second approach relies upon the sensitivity of G proteins to inactivation by NEM. Treatment of hippocampal slices with NEM blocks the modulatory effects of adenosine upon NE release from hippocampal slices, as well as the adenosine receptor mediated inhibition of adenylate cyclase (Fredholm and Lindgren, 1986, 1987); the $alpha_2$ receptor mediated inhibition of NE release is similarly affected (Allgaier et al, 1986). We have observed that NEM pretreatment can reduce the modulation of synaptic transmission at excitatory synapses in the CA1 region by adenosine as well (Dunwiddie and Fredholm, unpublished). Although this provides support for a G-protein involvement, the evidence is primarily correlative; NEM is a relatively non-specific agent, and there may be a variety of cellular proteins that are also affected under the same conditions that reduce responses to adenosine.

The final way in which the putative involvement of G proteins has been assessed has been through the use of pertussis toxin, which can ADP-ribosylate and inactivate specific G-proteins. This effect was initially established for G_i, but it is now apparent that other G proteins such as G_0 are substrates for this enzyme as well. The G proteins that couple adenosine receptors to potassium channels in hippocampal and striatal neurons are sensitive to the actions of pertussis toxin (Trussell and Jackson, 1986; Andrade et al, 1986; Dunwiddie and Proctor, unpublished) as are G proteins in other tissues such as the heart that couple A1 receptors to potassium channels (Bohm et al, 1986). The inhibition of LCB in hippocampal slices by adenosine is also completely abolished by pertussis toxin (Dunwiddie and Proctor, unpublished). The effects of pertussis toxin upon synaptic modulation are less consistent; in some systems, modulation of transmitter release is sensitive to pertussis toxin (e.g., $alpha_2$ inhibition of NE release; Allgaier et al, 1985). Dolphin and Prestwich (1985) reported that the inhibitory effect of adenosine on glutamate release from cultured cerebellar neurons was reversed by pertussis toxin treatment of the cultures. However, we have not observed any significant effect of pertussis toxin treatment on adenosine modulation of excitatory responses in hippocampus, even in slices in which the effect of adenosine on LCB is completely abolished (Dunwiddie, unpublished). One possible reason for such differences is that the modulatory effects in cerebellar cultures are mediated via adenosine receptors located upon the cell bodies or dendrites, and that the inhibition of release is due to a reduction in somatic excitability, as opposed to modulatory effects exerted directly upon the nerve terminals. The significance of results with pertussis toxin are made yet more difficult to interpret by the recent observation that there is a pertussis toxin sensitive G protein that can couple receptors to calcium channels as well (Holz et al, 1986; Scott and Dolphin, 1986).

At this point, the mechanism(s) that underlie the presynaptic modulation of excitatory transmission in the hippocampus remain unclear. Although there are at least 3 distinct postsynaptic actions of adenosine that, if transposed to the presynaptic nerve terminal, could account for diminished transmitter release, none of these actions appears to share all the properties of the presynaptic mechanism. The inhibition of calcium flux into the nerve terminal would be perhaps the simplest mechanism for the inhibition of transmitter release, but would not readily account for the observation that 4AP and cesium can block the actions of adenosine. A receptor coupled via a G-protein to a potassium channel, such as has been described in hippocampus and striatum, provides an alternative mechanism; however, this would not account for the apparent insensitivity of the presynaptic modulation to pertussis toxin, nor does this potassium channel appear to be sensitive to 4AP, as is the modulatory action of adenosine. The final postsynaptic effect, the inhibition of LCB by adenosine, is likely to involve some type of potassium channel, because it can be blocked by low concentrations of 4AP. Whatever the mechanism, it shares with presynaptic modulation the ability to be antagonized by 4AP but not TEA, and the pharmacological properties of the receptors involved are very similar; however, we have observed that in pertussis toxin treated slices, the effect of adenosine upon LCB but not the presynaptic effect is blocked. Thus, of the 3 putative mechanisms, each differs from presynaptic modulation in at least one major aspect.

Several hypotheses might be proposed to account for these differences. First, it is possible that the mechanism that underlies presynaptic modulation of release is entirely distinct from the mechanisms underlying postsynaptic actions of adenosine. Alternatively, there may be multiple mechanisms by which transmitter release is inhibited by adenosine. In this context, it is important to distinguish between treatments that have been reported to reduce the modulatory effects of adenosine, vs. those that completely block responses to adenosine. In the former case, this merely indicates that one of perhaps several contributory mechanisms have been blocked; a complete antagonism of the actions of adenosine suggests that either a single, specific mechanism is responsible for the effects of adenosine, or that the modulatory process has been blocked at a point prior to divergence into multiple actions. G proteins have been described that couple receptors to adenylate cyclase, potassium channels, calcium channels, and phosphoinositide turnover. Although the G proteins that mediate these effects are generally thought to be distinct but closely related in structure, it is entirely possible that a unique G protein, when activated by adenosine receptors, might possess the ability to interact with several such mechanisms to elicit a coordinated response.

In summary, it is clear that one could account for nearly all the presynaptic actions of adenosine based upon a mechanism involving either a reduction in a calcium conductance, an increase in potassium conductance, or both. However, the specific ways in which these cellular mechanisms operate to produce these responses, and the role played by other more complex cellular processes, such as cyclic AMP formation and subsequent protein phosphorylation, etc., remain to be determined. An attractive possibility is that adenosine might regulate activity both in the short term, via changes in membrane potential and calcium influx, and in the longer term via changes of a biochemical nature, such as phosphorylation of proteins in response to altered levels of calcium and cAMP.

REFERENCES

Allgaier, C., Feuerstein, T.J., Jackisch, R., and Hertting, G., 1985, Naunyn-Scmiedeberg's Archiv Pharmacology 331, 235.

Allgaier, C., Hertting, G. and Kugelgen, O.V., 1987, Br. J. Pharmacol., 90(2), 403-412.

Andrade, R., Malenka, R.C. and Nicoll, R.A., 1986, Society for Neuroscience Abstracts, 12, 15.

Bohm, M., Bruckner, R., Neumann, J., Schmitz, W., Scholz, H. and Starbatty, J., 1986, Naunyn Schmiedebergs Arch.Pharmacol. 332(4), 403-405.

Dolphin, A.C. and Archer, E.R., 1983, Neuroscience Letters, 43, 49-54.

Dolphin, A.C., Forda, S.R. and Scott, R.H., 1986, J. Physiol. 373, 47-61.

Dolphin, A.C., and Prestwich, S.A., 1985, Nature, 316, 148-150.

Dunwiddie, T.V., 1985, Int. Rev. Neurobiol. 27, 63-139.

Dunwiddie, T.V. and Fredholm, B.B., 1984, Naunyn-Schmiedeberg's Arch. Pharmacol. 326, 294-301.

Dunwiddie, T.V. and Fredholm, B.B., 1985, Adenosine modulation of synaptic responses in rat hippocampus: possible role of inhibition or activation of adenylate cyclase. In Advances in Cyclic Nucleotide and Protein Phosphorylation Research, 19, D.M.F. Cooper and K.B. Seamon (Eds.), 259-272.

Dunwiddie, T.V. and Hoffer, B.J., 1980, Br. J. Pharmacol., 69, 59-68.

Dunwiddie, T.V. and Proctor, W.R., 1987, Mechanisms underlying physiological responses to adenosine in the central nervous system. Topics and Perspectives in Adenosine Research, E. Gerlach and B. Becker, eds. (Springer Verlag), Berlin, p. 499.

Dunwiddie, T.V., Worth, T.S. and Olsson, R.A., 1986, Naunyn-Schmiedeberg's Arch. Pharmacol. 334, 77-85.

Fredholm, B.B. and Lindgren, E., 1986, Acta Physiol Scand, 126, 307-309.

Fredholm, B.B. and Lindgren, E., 1987, Acta Physiol Scand, 130(1), 95-105.

Greene, R.W., and Haas, H.L., 1985, J. Physiol. 366, 119-127.

Haas, H.L. and Greene, R.W., 1984, Pflugers Arch., 402, 244-247.

Haas, H.L., Jefferys, J.G., Slater, N.T. and Carpenter, D.O., 1984, Pflugers Arch., 400, 28-33.

Henon, B.K. and McAfee, D.A., 1983b, Modulation of calcium currents by adenosine receptors on mammalian sympathetic neurons. In: Regulatory function of adenosine, R. Berne, T. Rall and R. Rubio (Eds.), Martinus Nijhoff, The Hague, 455-406.

Hescheler, J., Rosenthal, W., Trautwein, W. and Schultz, G., 1987, Nature, 325, 445-446.

Holz, G.G., Rane, S.G. and Dunlap, K., 1986, Nature, 319, 670-672.

Kocsis, J.D., Eng, D.L. and Bhisitkul, R.B., 1984, Proc. Natl. Acad. Sci. (USA), 81, 6531-6534.

Lee, K.S., Schubert, P. and Heinemann, U., 1984, Brain Res., 321, 160-164.

Macdonald, R.L., Skerritt, J.H. and Werz, M.A., 1986, J. Physiol., 370, 75-90.

Madison, D.V., Fox, A.P. and Tsien, R.W., 1987, Biophysical Journal, 51, 30.

McCabe, J. and Scholfield, C.N., 1985, Pflugers Arch., 403, 141-145.

Miller, R.J., 1987, Science, 235, 46-52.

Okada, Y. and Kuroda Y., 1975, Proc. Jap. Acad., 51, 491-494.

Okada, Y. and Kuroda, Y., 1980, Euro. J. Pharmacol., 61, 137-146.

Okada, Y. and Ozawa, S., 1980, Euro. J. Pharmacol., 68, 483-492.

Perney, T.M., Hirning, L.D., Leeman, S.E. and Miller, R.J., 1986, Proc. Natl. Acad. Sci. (USA), 83, 6659.

Reddington, M., Lee, K.S. and Schubert, P., 1982, Neurosci Lett., 28, 275-279.

Scholfield, C.N., 1978, Brit. J. Phrmacol., 63, 239-244.

Scholfield, C.N., 1986, J. Physiol., 371, p. 100p.

Scholfield, C.N. and Steel, L., 1987, J. Physiol., (in press).

Schubert, P. and Lee, K.S., 1986, Neurosci Lett., 67, 334-338.

Schubert, P. and Mitzdorf, U., 1979, Brain Res., 172, 186-190.

Scott, R.H. and Dolphin, A.C., 1986, Neurosci Lett., 69(1), 59-64.

Segal, M., 1982, Euro. J. Pharmacol., 79, 193-199.

Silinsky, E.M., 1984, J. Physiol., 346, 243-256.

Silinsky, E.M., 1986, Trends in Pharmacol. Sci., 7, 180-185.

Silinsky, E.M. and Hirsh, J.K., and Vogel, S.M., 1987, Intracellular calcium mediating the actions of adenosine at neuromuscular junctions. In: Topics and Perspectives in Adenosine Research, edited by E. Gerlach and B. Becker (Springer-Verlag), Berlin, p. 537.

Stone, T.W., 1981, Neurosci., 6, 391-398.

Trussell, L.O. and Jackson, M.B., 1985, Proc. Natl. Acad. Sci. (USA), 82, 4857-4861.

Trussell, L.O. and Jackson, M.B., 1986, Society for Neuroscience Abstracts, 12, 15.

Relationship of pharmacological actions of adenosine to activation or inhibition of adenylate cyclase.

Bertil B. Fredholm, Eva Lindgren, Marianne Dunér-Engström, Johan Fastbom, James Wang[1], Johan Häggblad, Ingeborg van der Ploeg, Tommy Andersson[2], Mikael Jondal[3], Janet Ng and Christer Nordstedt.

Department of Pharmacology, Karolinska Institutet, Stockholm, Sweden, [1]Laboratory of Molecular and Cellular Neuroscience, The Rockefeller University, New York, USA, [2]Dept. Microbiology, Univ. Linköping, Linköping, Sweden, [3]Dept. Immunology, Karolinska Institutet, Stockholm, Sweden.

INTRODUCTION.

The adenosine receptors were subdivided into $A_1(R_i)$ and $A_2(R_s)$ primarily based on two criteria: relative agonist potency and on whether adenylate cyclase was inhibited or stimulated (Van Calker et al. 1979, Londos et al. 1980). It is important to consider whether all the effects of adenosine are *in fact* mediated via adenylate cyclase (cf. Fredholm 1982). In this paper we will briefly summarize some recent findings that indicate that the relationship between adenosine receptor stimulation and alterations in adenylate cyclase activity is not a simple one.

METHODS

$[^3H]$-transmitter release from rat hippocampus.

We used male Sprague-Dawley rats (150-250 g). The rats were killed by decapitation and the brains rapidly removed on ice. The hippocampus was dissected out and cut transversely into 0.4 mm thick slices using a manually operated McIlwain tissue chopper. After preincubation the slices were incubated with $[^3H]$-labelled transmitter or transmitter precursor in the presence of appropriate inhibitors of metabolic degradation (cf. Jonzon and Fredholm 1984). After washing, the slices were placed in chambers, superfused with gassed Krebs-Ringer bicarbonate buffer and stimulated with pulses of 10 V, 2 ms at the rate of 0.1-10 Hz via two silver electrodes. Two or three stimulations for 5 min were given and the fractional release of tritium during the second (S_2) or third (S_3) was divided by the amount released during the first (S_1); the basal outflow was subtracted.

Formation of [^{3}H] cyclic AMP

The tissues were preincubated and labelled with [^{3}H]-adenine in Krebs-Ringer bicarbonate buffer containing (except in the vas deferens experiments) rolipram 30 μM. The incubations were carried out at 37°C for 15 min with additions as indicated. At the end of the experiment the slices were homogenized and [^{3}H]-cyclic AMP was isolated. The results are expressed as per cent conversion of radioactivity to cyclic AMP, after correction for incomplete recovery.

[^{3}H]-inositolphosphate accumulation.

20 - 40 tissue prisms, or 10-100 10^{-6} cells, were preincubated in Krebs Ringer bicarbonate buffer for 30 min. They were then incubated in 3 ml fresh KRB, but with $CaCl_2$ reduced to 0.9 mM, containing 60 μl [^{3}H]-inositol (appr. 1 mC/ml) for a period of 2 - 5 hours. After washing 10 mM lithium chloride was added. The incubation time was 10 sec to 1 hr. The labelled inositol phosphates were separated essentially as described by Bone et al. (1984). The results are generally expressed as [^{3}H]- inositolphosphates in per cent of total water soluble [^{3}H]-inositol compounds.

Autoradiographic studies with [^{3}H]-CHA.

Sections of rat brain were labelled with [^{3}H]-N^6-cyclohexyladenosine (CHA) as previously described (Fastbom et al. 1986). In some experiments GTP was added. The tissues were exposed together with [^{3}H]-microscales (Amersham). The autoradiograms were analyzed using an IBAS image analysis system. Optical densities were transformed receptor densities by relating to the tissue equivalent values of the standards. Statistical hypotheses were tested with 1-way ANOVA.

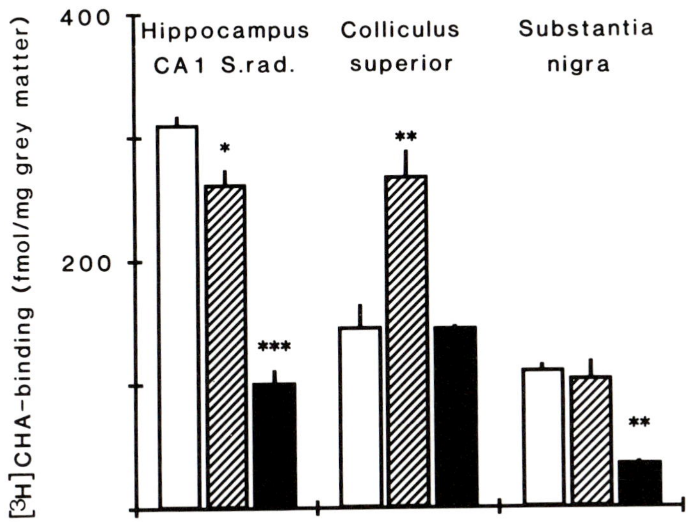

Fig.1. The effect of 0.5 and 5 μM GTP on the binding of [^{3}H]-CHA to three selected brain regions. Open bars: No GTP, Hatched bars :0.5 μM GTP, Filled bars: 5 μM GTP. * , ** , *** indicate significant differences: $p < 0.05$, $p < 0.01$ and $p < 0.001$, respectively.

RESULTS

Relationship of A_1-receptors and GTP-binding proteins.

There is excellent biochemical evidence that A_1-receptors are linked to GTP-binding proteins, as discussed in more detail by others at this meeting. We wanted to examine if quantitative autoradiography could be used to examine the possibility of regional heterogeneity in the coupling. As seen in Fig 1 we found that when low concentrations of GTP were used in the autoradiography the density of the [^{3}H]-CHA grains was slightly increased in some regions of the CNS, notably the superior colliculus, whereas they were reduced in other parts of the brain.

Further studies are required to asses the exact significance of this observation. It is however an intriguing possibility that it does represent a regional heterogeneity in the coupling between the A_1-receptors and the GTP-binding protein(s).

Presynaptic effects of adenosine.

Adenosine analogs are known to inhibit the release of several different transmitters. In the rat hippocampus we can find such inhibition of the release of $[^3H]$-noradrenaline (Jonzon and Fredholm 1984, Fredholm and Lindgren 1987), $[^3H]$-acetylcholine (Dunér-Engström and Fredholm 1987), $[^3H]$-glutamate (Fredholm et al. 1986), whereas the effect of adenosine on the release of GABA is limited or non-existent(unpublished). This could be related to the fact that adenosine is a better inhibitor of the release of excitatory than of inhibitory neurotransmitters (Snyder 1985).

There are several lines of circumstantial evidence that the presynaptic adenosine receptors are linked to a GTP-binding N-protein. 1. In many regions of the brain the adenosine receptors are definitely associated with presynaptic elements (cf. Snyder 1985). In all regions of the brain the binding of A_1-receptor ligands is strongly affected by GTP ($< 10\ \mu M$), indicating its association with N-proteins. 2. Using cultured cerebellar neurons Dolphin and Prestwich (1985) found that pertussis toxin blocked the presynaptic inhibitory effect of adenosine. 3. N-ethylmaleimide (NEM) is able to block N-proteins with some selectivity in hippocampus (Fredholm et al. 1985). Treatment with NEM was found to block $[^3H]$-NA, glutamate and ACh release in an apparently non-competitive manner (Allgaier et al. 1987, Fredholm et al. 1986, Dunér-Engström and Fredholm 1987, Fredholm and Lindgren 1987). While not conclusive these data strongly suggest a role for N-proteins in the presynaptic A_1-receptor effects.

By contrast, several lines of evidence indicate that the effect of adenosine on transmitter release is not (or at least not primarily) due to inhibition of adenylate cyclase activity. 1. The virtually complete inhibition by adenosine of transmitter release from some tissues implies that cyclic AMP would be required for transmitter release. This seems *a priori* unlikely (cf. Fredholm et al.1983). 2. Inhibitors of adenylate cyclase such as 2',5'-dideoxyadenosine do not reduce transmitter release (Fredholm and Lindgren 1987). 3. Increasing tissue cyclic AMP levels several fold with forskolin + a phosphodiesterase inhibitor does not significantly reduce the effect of adenosine analogs to block transmitter release (Fig 2, Fredholm et al. 1986, Dunér-Engström and Fredholm 1987, Fredholm and Lindgren 1987).

In order to determine which type of changes are produced in the nerve endings by A_1-receptor agonists we examined the effect of CHA on phosphorylation of Synapsin I. CHA was able to reduce, and delay, the depolarization-induced Ca^{2+}-dependent phosphorylation of a site on the Synapsin I molecule that is not a substrate for cyclic AMP dependent protein kinase. This is compatible with a role of adenosine to limit Ca^{2+}-entry.

It was not possible to consistently block or reduce the presynaptic adenosine effect by calcium-channel agonists (eg. BayK 8644) or antagonists (eg. nifedipine), that influence L-type calcium channels. There was some antagonism between BayK 8644 and R-PIA mediated inhibition of NA-release from the hippocampus, provided the release was evoked by electrical stimulation, but not if the release was evoked by potassium depolarization (Fredholm et al. 1986 and unpublished). $[^3H]$-ACh release was unaffected by BayK 8644 (Dunér-Engström and Fredholm 1987). Furthermore, there is some interaction between dihydropyridines and adenosine binding (Hu et al. 1987). Our interpretation is that dihydropyridine sensitive Ca-channels are not normally involved in electrically evoked transmitter release, but may become important if other channels, eg. N-channels (Miller 1987), are blocked by adenosine.

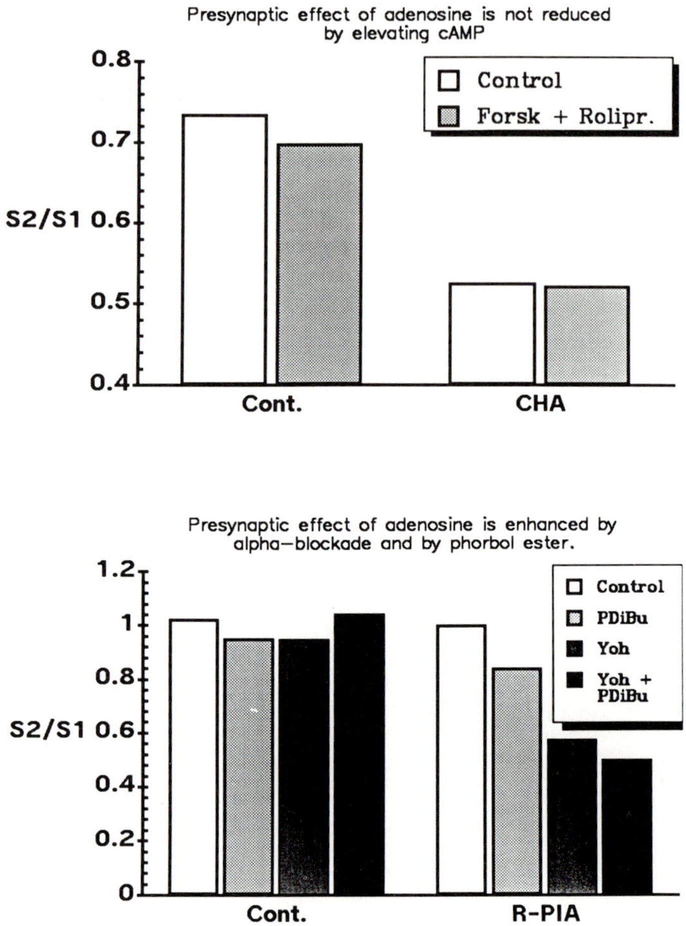

Fig 2. The effect of combined forskolin and rolipram (upper panel) and of phorbol 12,13-dibutyrate and/or yohimbine on the presynaptic effect of the adenosine analogues CHA (upper panel) and R-PIA (lower panel).

In agreement with others (eg. Wakade et al. 1985, Allgaier et al. 1986, Nichols et al. 1987) we found that phorbol esters, that stimulate protein kinase C (PkC, Nishizuka 1984), increased the release of [^{3}H]-NA from rat hippocampus. The increase amounted to between 2 and 3-fold.

It could not be blocked by nifedipine or by clonidine. However, as seen in Fig 2, the effect of R-PIA was if anything increased by treatment with phorbol dibutyrate. Thus, whereas α_2-adrenoceptor effects were blocked by phorbol ester, the effect of R-PIA was not. Differences between α_2-and other presynaptic receptors has been recognized before (Mulder and Schoffelmeer 1985). Comparisons with published results also indicate that there may be species differences. The conclusions from this series of experiments are summarized in Table 1.

Table 1. Summary of effects of adenosine on transmitter release.

1. The presynaptic inhibitory adenosine A_1-effects are probably mediated via a GTP-binding protein.
2. Inhibition of adenylate cyclase is usually not an important mechanism in inhibiting transmitter release.
3. The effect of adenosine analogs in nerve endings may involve a decreased entry of calcium.
4. The L-type voltage dependent calcium channel is not a major target for presynaptic adenosine receptors (even though dihydropyridines may block adenosine receptor binding).
5. Phorbol esters enhance transmitter release and, at least in the rat hippocampus, it blocks α_2-receptor mediated but not A_1-mediated presynaptic effects. The presynaptic adenosine action is enhanced by blocking alpha-autoreceptors.
6. There are differences between presynaptic agonists, between types of nerve endings, between species etc. in the relative importance of different mechanisms of inhibition of transmittor release. It is therefore probably very difficult to generalize from one system to another. It is also possible that presynaptic inhibitors such as adenosine can reduce transmitter release by several independent mechanisms.

Enhancement by adenosine of α_1-receptor mediated [^{3}H]-inositol phosphate accumulation.

In many tissues α_1-receptor stimulation leads to the accumulation of inositol phosphates (InsP). We found this to be true also in the rat vas deferens. On the other hand a stable ATP analogue, adenosine and the putative co-transmitter NPY had no effect on the accumulation of InsP's. However, as seen in Fig 3 there was a marked increase in the accumulation of phenylephrine-induced InsP accumulation when adenosine was added. This effect was shared by NPY and ATP. In parallel, adenosine decreased the accumulation of [^{3}H]-cyclic AMP in forskolin treated vasa deferentia. The two effects may be unrelated to each other, however, since forskolin, rolipram and dibuturyl cAMP did not significantly affect the accumulation of InsP induced by phenylephrine.

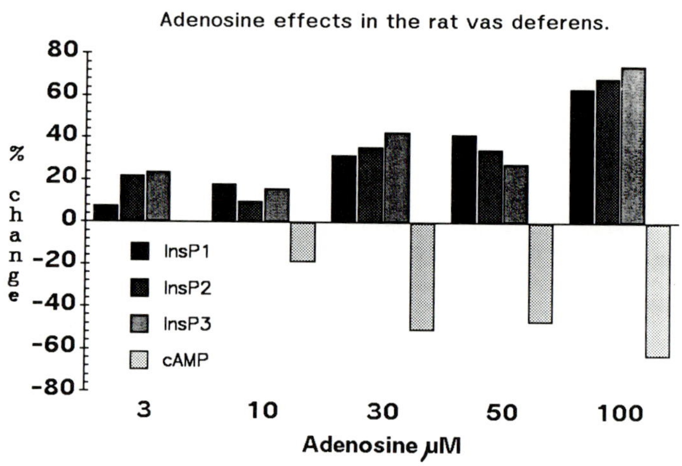

Fig 3. The effect of increasing concentrations of adenosine on the accumulation of inositol phosphates (InsP1, Insp2 and InsP3) stimulated by phenylephrine (50 μM) and the effect on [^{3}H]-cyclic AMP accumulation stimulated by 1 μM forskolin during a 5 min incubation.

<u>Enhancement of adenosine-mediated cyclic AMP accumulation by agents that activate PkC.</u>

It is well known that agents such as histamine (acting on H_1-receptors) and noradrenaline (acting on α_1-receptors) can increase the accumulation of cyclic AMP in brain slice preparations and that this is due to the enhancement of the actions of endogenous adenosine and/or prostaglandin (Daly 1977). These receptors are coupled to the breakdown of phosphatidyl inositol and the formation of diacylglycerol that can activate PkC. It was recently shown that this effect could be mimicked by tumor promoting phorbol esters (Hollingsworth et al. 1985, Nordstedt and Fredholm 1987). The mechanism behind this effect seemed to involve the coupling of the adenosine A_2-receptor to the adenylate cyclase rather than the receptor or the cyclase directly (Nordstedt and Fredholm 1987). We also found that the effect could not be accounted for by an inhibition of the inhibitory GTP-binding protein N_i.

Interestingly we have since repeated the basic finding that phorbolesters can stimulate adenosine receptor mediated cyclic AMP accumulation in a simpler system, the T-cell leukemia line Jurkat (Fredholm et al. 1987). In this cell activation of the T-cell receptors CD3 and CD2 by antibodies or lectins leads to a marked increase in the intracellular calcium concentration (Ng et al. 1987) and to an increased InsP formation. There is also evidence that PkC is activated. Interestingly both receptor-mediated stimulation of PkC by lectin and the addition of phorbol esters can potentiate the adenosine stimulated cyclic AMP accumulation in these cells.

Very recently we found that the effect of the phorbol ester differed depending upon the agent used to raise cyclic AMP accumulation. Not only adenosine, but also PGE_2, can stimulate cyclic AMP accumulation. Whereas the adenosine stimulated cyclic AMP accumulation was enhanced by tumor promoting phorbol esters the effect of PGE_2 was reduced. This in all probability means that the protein kinase C phosphorylates the receptor itself and in that way alters the effects produced by the agonist.

SUMMARY.

Much of the pioneering work on adenosine was done on cyclic AMP related systems- in many instances the very parameter studied was cyclic AMP. Therefore it is not surprising that changes in cyclic AMP (increases in the case of A_2-receptors or decreases in the case of A_1-receptors) was considered the major if not the only mechanism of action of adenosine acting on these receptors. It has even been proposed that one must postulate a third adenosine receptor A_3 because there are responses that may not be dependent on cyclic AMP (Ribeiro and Sebastiao 1986). This is obviously not warranted. In fact, the terminology A_1 and A_2 (rather than the alternative -R_i and R_a , which refer to changes in adenylate cyclase (Londos et al. 1980)) was adopted largely because these terms do not assume a particular mechanism of action.

Fig 4. The effect of phorbol dibutyrate (PDiBu, 0.1 μM) on the accumulation of [^{3}H]-cyclic AMP in Jurkat cells induced by adenosine (circles) or PGE_2 (squares). Filled symbols - no drug; open symbols - with PDiBu.

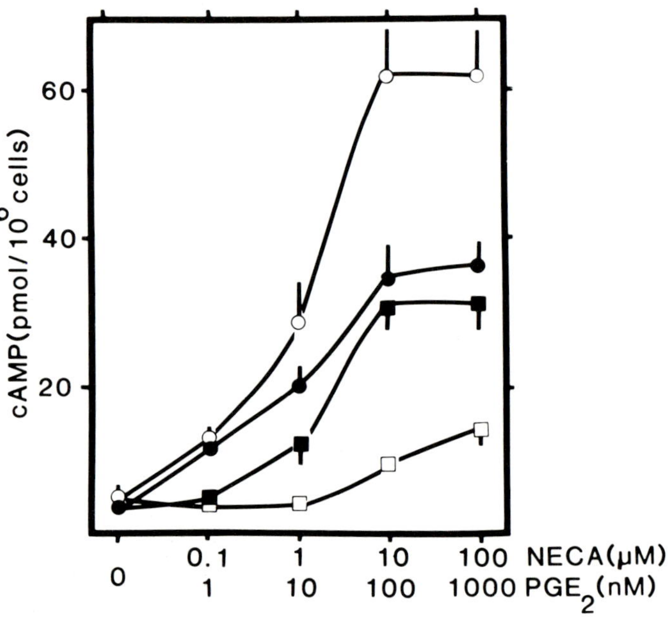

Here we have given further evidence that the relationship between adenosine receptor and adenylate cyclase is not a simple one. It seems possible that most adenosine receptors are linked to N-proteins. There is increasing evidence that there may be interactions between different N-proteins: The same receptor may interact with several N-proteins; several N-proteins may interact with one effector; the β,γ-subunits released from the activation of one N-protein may couple with the α-subunit of another and in that way weaken the signal transmitted by that α-subunit. These could be some of the reasons why an adenosine receptor that may primarily be linked to an ion channel causes inhibition of adenylate cyclase in brain. Similarly the potentiation by adenosine of the InsP formation that was observed in the vas deferens might be explained by an interaction between N-proteins.

Finally the results obtained indicate that the coupling between adenosine-receptors and cyclic AMP formation may be not only different between cells but can also vary in the same cell depending upon which other receptor pathways are simultaneously activated. This is in keeping with the opinion that a major role of adenosine is to act as a modulator of different hormones and transmitters.

REFERENCES.

Allgaier,C., Hertting, G. Br J Pharmacol 90, 403.

Allgaier,C., Von Kugelen,O. Europ J Pharmacol, 129, 389.

Bone, E.A.,Fretten, P., Palmer, S. ,Kirk,C.J. & Michell,R.H., 1984, Biochem J, 221, 803-811.

Daly, J.W.: Cyclic nucleotides in the nervous system. Plenum Press, New York, 1977.

Dolphin, A.C. & Prestwich,S.A.,1985, Nature, 316, 148.

Dunér-Engström, M. & Fredholm, B.B., 1987, Submitted to J.Neurochem.

Fastbom, J., Pazos, A., Probst, A., & Palacios, J.M., 1986, Neurosci lett., 65, 127.

Fredholm, B.B., 1982, Med Biol 60, 289.

Fredholm, B.B., Fastbom, J. & Lindgren, E., 1986, Acta Physiol Scand 127, 381.

Fredholm, B.B., Gustafsson, L., Hedqvist, P. & Sollevi, A. 1983. Adenosine in the regulation of neurotransmitter release in the peripheral nervous system. In: R.M. Berne, T.V. Rall & R. Rubio (eds) Regulatory Function of Adenosine,(Martinus Nijhoff Publishers) p. 479.

Fredholm, B.B.,Hu, P.S. & Lindgren,E.,1986, Acta Physiol. Scand. 1986, 128: 659-660.

Fredholm, B.B., Jondal, M. & Nordstedt, C., 1987, Biochem Biophys Res Commun, 145, 344.

Fredholm, B.B. & Lindgren, E., 1987, Acta physiol scand. 130, 95.

Fredholm, B.B., Lindgren, E. & Lindström, K. 1985, Br J Pharmacol 86, 509.

Hollingsworth, E.B., Sears,E.B. & Daly,J.W., 1985, FEBS Lett 184, 339.

Hu, P.-S., Lindgren, E., Jacobson, K.A. & Fredholm, B.B., 1987, Pharmacol & Toxicol. 61, 1.

Jonzon, B., & Fredholm, B.B.,1984, Life Sci 35, 1971.

Londos, C., Cooper, D.M. & Wolff, J.,1980, Proc Natl Acad Sci USA 77,2551.

Miller, R.J.,1987, Science 235, 46.

Mulder, A.H. & Schoffelmeer, A.N.M., 1985, Adv Cycl Nucl Res 19, 273.

Nichols, R.A., Haycock, J.W., Wang, J.K.T. & Greengard, P., 1987, J Neurochem 48, 615.

Nishizuka, Y.,1984, Nature , 308, 693.

Ng, J., Fredholm, B., Jondal, M. & Andersson, T., 1987 Int J Immunopharm 9, 17.

Nordstedt, C. & Fredholm, B.B.,1987, Naunyn-Schmiedeberg's Arch Pharmacol 335, 136.

Ribeiro, J.A., & Sebastiao, A.M., 1986, Progr. Neurobiol 26, 179.

Snyder, S.H., 1985, Ann. Rev. Neurosci 8, 103.

VanCalker, D., Muller, M. & Hamprecht, B., 1979, J Neurochem 33, 999.

Wakade, A.R., Malhotra, R.K. & Wakade, T.D., 1985, Naunyn-Schmiedeberg's Arch Pharmacol 331, 122.

Acknowledgements. We thank Dr. Paul Greengard for the facilities used in the phosphorylation experiments and for helpful discussions. These studies were supported in part by the swedish medical research council (no. 2553), Ostermans foundations, Stiftelsen Gamla tjänarinnor, Astra ALAB, and Karolinska Institutet.

ACTIONS OF ADENOSINE ON CENTRAL NEURONES

Penelope A. Brooks, J. T. Bartrup & T. W. Stone

Dept. of Physiology, St. George´s Medical School,
University of London, London SW 17 ORE.

INTRODUCTION

A number of actions of adenosine and other purines on neurones are now well established. On both peripheral and central neurones there are receptors mediating the inhibition of transmitter release, these receptors having a number of features resembling the A1 population of receptors responsible for inhibiting adenylate cyclase. In some cases there is also evidence for a stimulant action of purines on transmitter release, for which the receptors more nearly resemble the cyclase linked A2 population (Spignoli et al., 1984; Silinsky, 1980). These actions have been clearly demonstrated on cholinergic and monoaminergic releasing terminals (Fredholm & Hedqvist, 1980; Stone, 1981), and may involve changes of calcium influx or an interference with intraterminal calcium availability (Silinsky, 1984; 1985).

Whereas these effects on transmitter release are seen at relatively low concentrations of purines, for example low micromolar adenosine, higher concentrations can have direct hyperpolarizing effects on neuronal somata. In hippocampal pyramidal cells for instance adenosine at concentrations greater than about 10 uM produces hyperpolarisation when applied directly to the cell soma by microiontophoresis (Segal, 1982; Siggins & Schubert, 1981).

Modulation of receptor sensitivity

In addition to these effects of adenosine acting alone there are increasing numbers of examples of purines modifying the sensitivity of cells to classical transmitters. Hedqvist & Fredholm (1976), Kazic &

Milosavljevic (1980), Holck & Marks (1970) and Kennedy & Burnstock (1986) have described changes of noradrenaline sensitivity of smooth muscle preparations by adenosine. Long & Stone (1986) have reported changes in the resensitisation of rat vas deferens by purines, which appear to involve alpha 2-adrenoceptors and may thus be related to the biochemical studies by Watanabe et al. (1983) in which incubation of the vas deferens with 2-chloroadenosine induced previously undetectable alpha-2 receptor binding.

There are also indications that purines can act to modify the sensitivity of neurones to acetylcholine. Evidence for such an effect was obtained by Ewald (1976) with the observation that ATP could potentiate end-plate depolarisation produced by acetylcholine. Subsequently Akasu and his colleagues (Akasu et al., 1981; Akasu & Koketsu, 1985) demonstrated that this, and a related action on nicotinic receptors of bullfrog sympathetic ganglion cells, could not be attributed to a change in receptor properties or kinetics, and that the potentiating effect of adenosine was probably due to an increase in the number of acetylcholine activated ionic channels or to an increase of single channel conductance.

Less work has been performed on possible purine modulation of cholinergic muscarinic receptors, though Gustaffson (1981) did observe an enhancement by adenosine of acetylcholine responses in the rabbit gastrointestinal system and Cox & Walker (1985) have observed changes in the muscarinic sensitivity of Helix neurones.

Muscarinic sensitivity in the CNS

We have therefore recently begun to examine the interactions between adenosine and carbachol or amino acids in the rat hippocampal slice preparation. Since adenosine and its analogues are such effective inhibitors of synaptic transmission, the experiments were performed primarily using antidromically evoked population spikes in the CA1 pyramidal cell layer as a measure of neuronal excitability. Carbachol was applied into this cell layer using microiontophoresis (Stone, 1985). Pulses of carbachol were effected at regular intervals using currents of 5 to 100 nA for 5 to 25 seconds. This produced a depression of the population spike size which was presumed to reflect depolarisation of the cells. This assumption was based on the fact that previous studies of muscarinic activity on these cells had revealed only depolarising activity (Dodd & Kelly, 1981; Benardo & Prince, 1982a, b; Halliwell & Adams,

1982) and that subsequent single cell analyses of our own also detected only increases of neuronal firing rate. Responses to carbachol in our experiments were blocked by atropine 0.1 uM.

A section of a typical experiment is illustrated in Fig. 1. During the addition of adenosine to the medium superfusing the brain slices the responses to carbachol were substantially inhibited, though it is important to note that in most experiments there was no change in the size of the baseline population spike. Dose response curves were produced for this phenomenon and for the similar activity shown by a series of adenosine analogues including R-phenylisopropyladenosine (R-PIA), and 5´N-ethylcarboxamide adenosine (NECA). Both these analogues were significantly more active than adenosine itself, though similar to each other in potency.

ATP was also able to induce a comparable suppression of carbachol sensitivity, but at concentrations equivalent to those at which adenosine was highly active, ATP was much less effective. In addition 8-phenyltheophylline prevented completely these actions of both adenosine and ATP, implying that ATP was only acting after metabolism to adenosine, at a Pl class of purine receptor (Burnstock, 1978). Preliminary experiments using the more stable analogues of ATP appear to confirm this conclusion, since their activity is weak, and can still be blocked by 8-phenyltheophylline.

The pharmacological specificity of these actions of adenosine and its analogues is suggested by the fact that there was normally no change of the antidromic spike size in the presence of the purines alone, and the fact that adenosine, for example, was active at concentrations which have previously been found to produce little or no change of neuronal resting membrane potential (Siggins & Schubert, 1981). Nevertheless, in order to obtain more information on the mechanism of the interaction we have compared the effect of adenosine on responses of pyramidal neurones to excitatory amino acids. In studies of population spike amplitude we find that adenosine does not change at all the depression of the potential produced by quisqualic acid, at concentrations of 100 uM, which abolish sensitivity to carbachol.

Single cell studies.

In separate experiments on single cell activity, we have compared the excitatory activity of carbachol, quisqualic acid, kainic acid and N-methyl-D-aspartic acid

Fig. 1 Record of the amplitude of the CA1 antidromic population spike, which is reduced by iontophoretic applications of carbachol (C) 35 nA for 25 seconds. Perfusion of the hippocampal slice with medium containing adenosine, 100 uM abolishes the sensitivity to carbachol. In parallel experiments, sensitivity to quisqualic or kainic acids was little affected by adenosine.

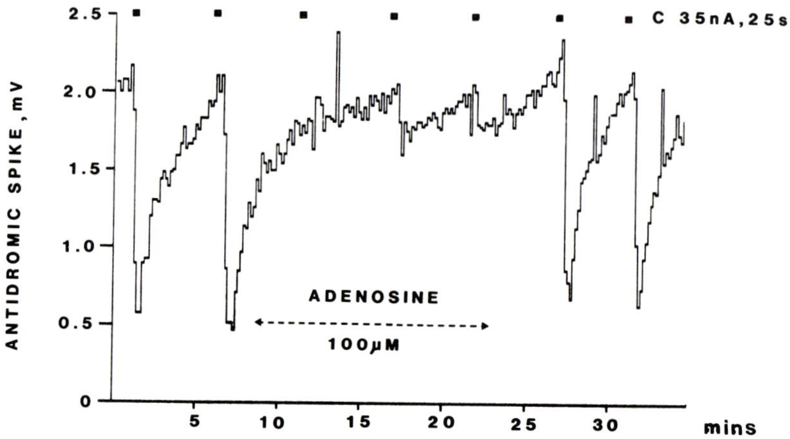

(NMDA), selective agonists for three postulated subtypes of acidic amino acid receptor in the central nervous system (Watkins & Evans, 1981).

The absence of any modification by adenosine of quisqualate responses was confirmed. Similarly at a concentration of 10 uM adenosine, which normally suppressed carbachol excitation, there was no significant change in responses to kainic acid. Some reduction was noted in sensitivity to NMDA, however. This could conceivably result from a very small change of neuronal excitability which could have a disproportionately large inhibitory effect on NMDA sensitivity by virtue of the voltage dependent nature of NMDA activated channel blockade by magnesium (Nowak et al., 1984; Mayer & Westbrook, 1987).

Since magnesium has also been said to be an essential link in the activation of adenylate cyclase by adenosine (Yeung et al., 1985) experiments were also performed to assess the relevance of this ion to the various actions of adenosine in hippocampal slices. In normal magnesium (1.2

mM) solutions adenosine produced an inhibition of orthodromically evoked synaptic potentials with an ED50 of 6.8 uM. In nominally magnesium free media, however, this action became very weak, the ED50 being difficult to ascertain. At a concentration of 100 uM for instance adenosine was now able to produce only an approximately 15% inhibition of evoked population spikes. When tested on excitatory responses to iontophoretically applied agents as outlined above, however, adenosine was able to suppress carbachol (and to a lesser extent NMDA) sensitivity approximately as effectively in the presence or absence of magnesium. While these latter experiments do not therefore assist greatly in understanding the site and mechanism of the interactions reported here, they do serve to emphasise the difference between these effects and the frequently studied depression of transmitter release.

Further studies are in progress to attempt to clarify the mechanism involved, which may be a modulation of a common potassium channel. Muscarinic agonists are known to excite pyramidal cells by changing potassium conductances and a similar explanation has been proposed for the effects of adenosine on a variety of cell membranes (Segal, 1982; Lotan et al., 1982).

Future significance

The potency and selectivity of these effects may have major implications both for appreciating the specific neurochemical substrate for the CNS stimulant actions of xanthines, and for the future development of compounds of potential use in disorders where cholinergic underactivity may be a relevant feature, such ad Alzheimer´s disease.

ACKNOWLEDGEMENTS

We are grateful to the Wellcome Trust for supporting this work. J. T. Bartrup is an S. E. R. C. research student.

REFERENCES

Akasu, T., Hirai, K. & Koketsu, K., 1981, British Journal of Pharmacology, 74, 505.

Akasu, T. & Koketsu, K., 1985, British Journal of Pharmacology, 84, 525.

Benardo, L. S. & Prince, D. A., 1982a, Brain Research, 249, 315.

Benardo, L. S. & Prince, D. A., 1982b, Brain Research, 249,333.

Burnstock, G., 1978, In Cell Membrane Receptors for Drugs and Hormones: A Multidisciplinary Approach, eds. R. W. Straub & L. Bolis, (Raven Press, N. Y.) p.107.

Cox, R. T. L. & Walker, R. J., 1985, British Journal of Pharmacology, 86, 449p.

Dodd, J. & Kelly, J. S., 1981, Brain Research, 205, 337.

Ewald, D. A., 1976, Journal of membrane biology, 29, 47.

Fredholm, B. B. & Hedqvist, P., 1980, Biochemical Pharmacology 29, 1635.

Halliwell, J. V. & Adams, P. R., 1982, Brain Research, 250, 71

Hedqvist, P. & Fredholm, B. B., 1976, Archives of Pharmacology 293, 217.

Gustaffson, L., 1981, Acta Physiologica Scandinavica 111, 263.

Holck, M. H. & Marks, B. H., 1978, Journal of Pharmacology and Experimental Therapeutics, 205, 104

Kazic, T. L. & Milosavljevic, D., 1977, Journal of Pharmacy and Pharmacology, 29, 542.

Kennedy, C. & Burnstock, G., 1986, Journal of Pharmacy and Pharmacology, 38, 307.

Long, C. J. & Stone, T. W., 1986, European Journal of Pharmacology, 132, 11.

Lotan, I., Dascal, N., Cohen, S. & Lass, Y., 1982, Nature, 298, 572.

Mayer, M. L. & Westbrook, G. L., 1987, Progress in Neurobiology, 28, 197.

Nowak, L., Bregestovski, P., Ascher, P., Herbert, A. & Prochiantz, A., 1984, Nature, 307, 462.

Segal, M., 1982, European Journal of Pharmacology, 79, 193.

Siggins, G. R. & Schubert, P., 1981, Neuroscience Letters, 23, 55.

Silinsky, E. M., 1980, British Journal of Pharmacology, 71, 191.

Silinsky, E. M., 1984, Journal of Physiology, 346, 243.

Silinsky, E. M., 1985, In Purines: Pharmacology and Physiological Roles, ed. T. W. Stone, (Macmillan Press) p. 67.

Spignoli, G., Pedata, F. & Pepeu, G., 1984, Euopean Journal of Pharmacology, 97, 341.

Stone, T. W., 1981, Neuroscience, 6, 523.

Stone, T. W., 1985, Microiontophoresis and Pressure Ejection, (Wiley & Sons).

Watanabe, Y., Rong-Tsan, L. & Yoshida, H., 1983, European Journal of Pharmacology, 86, 265.

Yeung, S.-M. H.,Frame, L. T. & Venter, L. C. 1985, In Purines: Pharmacology & Physiological Roles, ed. T. W. Stone (Macmillan Press, p.203).

ACTIONS OF ADENOSINE ON THE CARDIOVASCULAR SYSTEM

CARDIAC ELECTROPHYSIOLOGIC ACTIONS OF ADENOSINE AND ADENOSINE 5'-TRIPHOSPHATE

Amir Pelleg

The Lankenau Medical Research Center and Medical College of Pennsylvania, Philadelphia, Pennsylvania, USA

Introduction

The purine nucleoside adenosine and its related nucleotide adenosine 5'-triphosphate (ATP) are two biological compounds which are found in every cell of the human body. In 1929, Drury & Szent-Gyorgyi reported for the first time on the actions of adenosine and related compounds on the cardiovascular system. They suggested that these compounds, in addition to their metabolic role, have physiological activity which could be "involved in cell function not only in the heart but in the body generally" (Drury & Szent-Gyorgyi, 1929). The observations reported in their classical paper, which have been confirmed in numerous studies in the past six decades (for recent reviews see, Belardinelli et al. 1983; Belhassen and Pelleg, 1984; Pelleg, 1985), have laid the foundation to the field of cardiac electrophysiology and pharmacology of adenosine and ATP.

This brief review updates information with regard to recent developments in the field and calls attention to the major unresolved questions.

Electrophysiologic Effects of Adenosine and ATP

The most pronounced effects of adenosine and ATP are negative chronotropic effects on automatic activity of sinoatrial node (SN), atrioventricular (AV) junction and ventricular pacemakers, as well as negative dromotropic effect on AV conduction. Since the early studies of Drury & Szent-Gyorgyi (1929), these effects have been shown both in vitro and in vivo in wide variety of species including rat, guinea pig, rabbit, cat, dog, monkey and human.

While the effects of adenosine and ATP in different species are similar, there is species variability with regard to certain aspects of their actions. For example, the guinea pig AV node is more sensitive to adenosine than that of the rat. In addition, in the human and guinea pig heart the AV node is more sensitive to adenosine than the SN and therefore, the negative dromotropic action of the nucleoside is more pronounced than its chronotropic action. However, in the canine heart the SN is more sensitive than the AV node. Also, ATP can trigger a vagal reflex in the feline, canine and human but not in the rodent and lagomorpha heart (for discussion see Pelleg, 1987). This reflex, when operative, is responsible for the differential potencies of adenosine and ATP actions on sinus and AV nodes (Pelleg et al., 1985a; 1985b) but does not play a major role in the negative chronotropic action of ATP on ventricular pacemakers (Pelleg et al., 1986b). While adenosine does not trigger a vagal reflex similar to ATP, there is evidence for adenosine-vagal interaction which could modulate adenosine's electrophysiological effects (see below: Mechanisms of Action). In addition, the sensitivity of the canine AV node to ATP is much higher than to adenosine (apparently independent of the autonomic nervous system).

Another aspect of species variability is the effect of dipyridamole on adenosine's actions. Dipyridamole an inhibitor of adenosine transport into cells (Kolassa et al., 1970) potentiated the action of adenosine in the guinea pig but not rat heart (Stafford, 1966; Belardinelli et al., 1982).

Finally, species variability has recently been observed also with regard to the indirect action of adenosine (see below) (Hopwood et al., 1987).

Electropharmacology of Adenosine and ATP

The negative dromotropic action of adenosine and ATP on AV nodal conduction has led to the use of these compounds as antiarrhythmic agents for the acute management of paroxysmal reentrant supraventricular tachycardia involving the AV node (see Belhassen & Pelleg, 1984; Belhassen & Pelleg, 1985; Munoz et al., 1987; DiMarco, 1987).

A recent report has described exercise-induced ventricular tachycardia in humans sensitive to adenosine (Lerman et al., 1986). Earlier observations in experimental models of ventricular arrhythmias have suggested antiarrhythmic activity of adenosine mediated by local or central action of the nucleoside (Fagbemi & Parratt, 1984; Huang & Yang, 1985; Cuparencu et al., 1986; Laorden et al., 1986). Similar antiarrhythmic activity of ATP have also been documented (Sohn et al., 1982; Laorden et al., 1986).

The exact mechanisms of this antiarrhythmic action is

not known. However, the anti-adrenergic activity of adenosine at the pre- and post-synaptic level (see Pelleg, 1987), effects on Ca^{2+} influx (Schrader et al., 1975), increased blood flow to the ischemic myocardium, and effects on purine metabolism and energy preservation (VanBelle, 1985) could be involved in these beneficial effects of adenosine and ATP.

Adenosine and ATP interact with different drugs and this interaction could modulate their electrophysiologic actions. Cardiac glycosides were found to poteniate the actions of adenosine (Rand et al., 1955) probably due to inhibition of adenosine deaminase (Rand et al., 1955). However, in a small sinus nodal preparation West & Belardinelli (1985) found that ouabain caused slight and nonsignificant enhancement of the hyperpolarizing effect of adenosine. Similar results were obtained in the frog sinus venosus (Hartzell, 1979). Methodological differences are probably responsible for these inconsistent results.

Several vasodilators as well as drugs with calcium channel blocking activity have been shown to potentiate the actions of adenosine. These include: dipyridamole (as mentioned above), papaverine, dilazep, lidoflazine and benzodiazepines (for references see discussion in Pelleg et al., 1987d). Verapamil, which was shown to inhibit the binding of adenosine to uptake sites (Marangos et al., 1984) and adenosine transport (Ford et al., 1985), potentiated the electrophysiological actions of adenosine both in vitro (Meszaros et al., 1986) and in vivo (Pelleg et al., 1987d). However, in the latter case this action was probably mediated by the autonomic nervous system (Pelleg et al., 1987d). In the spontaneously beating atria of the guinea pig, small doses of verapamil did not affect the negative chronotropic action of adenosine (Fujita et al., 1980).

Indomethacin, which inhibits adenosine uptake (Phillis & Wu, 1981) was also found to potentiate the negative chronotropic action of adenosine (Pelleg et al., 1986a). Finally, the Class I agent, quinidine, diminished the electrical and mechanical changes in guinea pig atria treated with adenosine (Meszaros et al., 1986).

Arrhythmogenic Effect of Adenosine and ATP

Adenosine and ATP are released from cardiac cells in increased amounts during ischemia or hypoxia. In 1965, James speculated that "sinus bradycardia and sinus arrest which are sometimes observed during myocardial infarction may be partially mediated through the action of ATP (or adenosine) on the node." This hypothesis is supported by more recent findings (Baldauf, 1971; Szentmiklosi et al., 1979; Motomura & Hashimoto, 1982; Wesley et al., 1986). With regard to the AV node Belardinelli et al. (1981) concluded that they "have demonstrated that adenosine fulfills some of the important criteria that any candidate must meet to be postulated as a mediator of the ischemic A-V block." Further studies by Belardinelli and his colleagues as well as others have established the role of endogenous adenosine in AV conduction disturbances under pathophysiologic conditions (Belardinelli et al., 1980; Belardinelli et al., 1982; Watanabe et al., 1983; Clemo & Belardinelli, 1986a; 1986b).

Recent observations in the clinical setting have supported the voluminous experimental data suggesting mediation by endogenous adenosine of the ischemia induced AV nodal conduction disturbances (Wesley Jr. et al., 1986; Shah et al. 1987).

Atrial arrhythmias mediated by acetylcholine (or vagal input) are well documented. Thus, in view of the similarity between the mechanism of action of acetylcholine and adenosine and ATP (see below, Mechanisms of Action) the reports on atrial arrhythmias induced by exogenous adenosine and ATP are not surprising. Both compounds were reported to induce atrial flutter/fibrillation in the human and canine heart (DiMarco et al., 1983; Belhassen et al., 1984; Munoz et al., 1985).

Mechanisms of Action

In sinus and AV nodal cells as well as in atrial myocardial cells adenosine causes hyperpolarization associated with increased rate of rise of the action potential and decrease in rate of phase four depolarization (see Pelleg, 1985). These effects are mediated by increased potassium conductance (Belardinelli & Isenberg, 1983; Jochem & Nawrath, 1983; Meszaros et al., 1984) similar to the electrophysiologic actions of acetylcholine (Kurachi et al., 1986).

While Urthaler et al. (1981) reported that adenosine significantly reduced the action potential rate of rise of paced canine right ventricular trabeculae, other studies failed to detect any significant effect of adenosine on ventricular myocardial cells (see Pelleg, 1985; Solti et al., 1984). In contrast, adenosine does affect pacemaker activity of Purkinje fibers. Adenosine decreased the rate of phase four depolarization in guinea pig Purkinje fibers and attenuated the automaticity of canine Purkinje fibers (see Pelleg, 1985).

These actions of adenosine are mediated by specific

receptors. Burnstock (1980) was first to propose that purinergic receptors P_1 and P_2 mediate the cardiac actions of adenosine and ATP respectively. Several studies have shown that the negative chronotropic and dromotropic actions of adenosine are mediated by specific extracellular receptors (Paton & Kurahashi, 1981; Belardinelli et al., 1982). The actions of adenosine at these receptor sites are competitively inhibited by xanthine derivatives and potentiated by adenosine transport blockers (Stafford, 1966).

So far there is no evidence to support the presence of P_2-purinergic receptors in the mammalian heart which mediate specifically the electrophysiologic actions of ATP. On the contrary, several studies have indicated that the action of ATP is due to its breakdown to adenosine when its interaction with the autonomic nervous system is prevented (Belardinelli et al., 1984; Pelleg et al., 1987a). However, direct action of ATP at P_1-purinoceptor sites without degradation to adenosine cannot be excluded (Collis & Pettinger, 1982). ATP can trigger a strong vagal reflex in several species. The degree to which this reflex plays in mediating the actions of ATP is dependent on the mode of its administration (for a detailed discussion see Pelleg, 1987). Thus, several interventions which enhance vagal effects on the heart can also potentiate the actions of ATP including the administration of physostigmine (Wayne et al., 1949; Pelleg et al., 1985b) and elevated plasma levels of Ca^{2+} (Pelleg et al., 1987b). It should be noted that while Ca^{2+} did not potentiate the action of adenosine (Drury & Szent-Gyorgyi, 1929; Chiba et al., 1973; Pelleg et al., 1987b), physostigmine did cause small but significant enhancement of the negative chronotropic action of adenosine (Pelleg et al., 1985b). The latter was probably due, at least in part, to the complex interaction between adenosine and vagal input to the heart (Pelleg et al., 1987c), which under certain conditions, can result in potentiation of adenosine's effects (Mazgalev et al., 1987; Pelleg et al., 1987c).

Both adenosine and ATP can trigger the release of prostaglandins in the heart (Needleman et al., 1974; Zehl et al., 1976), however, there is no evidence for the involvement of prostaglandins in the electrophysiologic actions of adenosine (Pelleg et al., 1986a).

Numerous studies in vitro have shown that in addition to its direct action adenosine can exert pronounced effect indirectly by antagonizing the effects of catecholamines (for references and discussion see Pelleg, 1987). However, attempts to demonstrate the indirect action of adenosine in vivo have been unsuccessful (Seitelberger et al., 1984; Raberger et al., 1987). It has been recently suggested that even under physiological conditions (characterized by relatively low levels of extracellular adenosine) adenosine "appears to serve an important physiological function as a negative feedback modulator of β -adrenoceptor-mediated contractile and metabolic responses" (Dobson et al., 1987). This attractive hypothesis is in discordant with the hypothesis of Schrader et al. according to which "the first response of the heart to increased adenosine formation is vasodilation. Only when the coronary reserve is exhausted (either by maximum vasodilation or coronary stenosis) does the antiadrenergic action of adenosine come into play." (See discussion following Raberger et al., 1987, p. 391). Thus, the physiological role of catecholamine antagonism by endogenous adenosine is still undetermined. However, as mentioned earlier, recent data obtained in the clinical setting was interpreted to suggest catecholamine antagonism by exogenous adenosine in the human heart (Lerman et al., 1986).

The exact post receptor reaction cascade which mediates the electrophysiologic effects of adenosine is not fully known. One of the major controversial questions is the involvement of cyclic adenosine 3', 5'-monophosphate (cAMP)in the actions of adenosine (see Pelleg, 1987). Recently it has been suggested that adenosine can increase the rate of inositol lipid metabolism which leads to the formation of inositol 1,4,5 triphosphate (IP_3) and the subsequent release of Ca^{2+} from intracellular stores (Rubio et al., 1986). This provocative hypothesis which designates Ca^{2+}, phospholipid and phospholipid dependent Ca^{2+}-activated protein kinase as second messengers of adenosine is supported by (a) the ability of adenosine to exert a positive inotropic action under certain experimental conditions (Masse et al., 1986; Rubio et al., 1986), (b) the increase by adenosine of the residual and strontium extracted Ca^{2+} in the carp atrium (Rotmensch et al., 1982) and (c) the increase of free calcium in endothelial cells caused by ATP and related compounds (Luckhoff & Busse, 1986). In addition, acetylcholine, whose cardiac actions are closely resembled by adenosine, exerted a positive inotropic effect in canine cardiac Purkinje fibers (Gilmour & Zipes, 1985). Furthermore, stimulation of cardiac muscarinic cholinergic receptors resulted in stimulation of inositol phosphate formation (Brown & Jones, 1986; Quist & Satumtira, 1987). This action is apparently not inhibited by pertussis toxin (Masters et al., 1985) which attenuates cAMP mediated actions of adenosine and acetylcholine.

A different conclusion was reached by Kurachi et al. (1986) based on their data in isolated guinea pig atrial cells. Using "inside out" patch conditions Kurachi et al. (1986) demonstrated that GTP, but neither cAMP or change in intracellular Ca^{2+}, was the only requirement for the activation of K^+ channels by adenosine and acetylcholine. Thus, these authors concluded that "purinergic and muscarinic activation of K^+ channel in cardiac cell membrane does not require intracellular second messengers, such as cyclic nucleotides and Ca^{2+} (Kurachi et al., 1986). This conclusion is supported by the finding that potent adenosine analogs were ineffective at inducing inositol phosphate formation (see Brown & Jones, 1986). It should be noted here that the differential action of acetylcholine (and probably also that of adenosine) in the atrial and ventricular tissues do not appear to be due to differences in the GTP-binding proteins which mediate the actions of acetylcholine in the avian heart (Martin et al., 1986).

Thus, the elucidation of post adenosine receptor reaction cascade which mediates the electrophysiologic actions of adenosine awaits further studies.

References

Baldauf, J., 1971. Release of a negative chronotropic and dromotropic substance during cardic hypoxia. (Abstract) Acta Pharmacologica Toxicologica, 29, 124.

Belardinelli, L., Belloni, F.L., Rubio, R. & Berne, R.M., 1980. Atrioventricular conduction disturbances during hypoxia. Possible role of adenosine in rabbit and guinea pig heart. Circulation Research, 47, 684.

Belardinelli, L., Mattos, E.C. & Berne, R.M., 1981. Evidence for adenosine mediation of atrioventricular block in the ischemic canine myocardium. Journal of Clinical Investigation, 68, 195.

Belardinelli, L., Fenton, R.A., West, G.A., Linden, J., Althaus, J.S. & Berne, R.M., 1982. Extracellular action of adenosine and antagonism by aminophylline on the atrioventricular conduction of isolated perfused guinea pig and rat hearts. Circulation Research, 51, 569.

Belardinelli, L., West, G.A., Crampton, R. & Berne, R.M., 1983. Chronotropic and dromotropic effects of adenosine. In Regulatory Function of Adenosine, edited by R.M. Berne, T.W. Rall & R. Rubio (Martinus Nijhoff, Boston), p. 377.

Belardinelli, L. & Isenberg, G., 1983. Isolated atrial myocytes: Adenosine and acetylcholine increase potassium conductance. American Journal of Physiology, 244, H734.

Belardinelli, L., Shryock, J., West, G.A., Clemo, H.F.,

DiMarco, J.P. & Berne, R.M., 1984. Effects of adenosine and adenine nucleotides on the atrioventricular node of isolated guinea pig hearts. Circulation, 70, 1083.

Belhassen, B. & Pelleg, A., 1984. Electrophysiologic effects of adenosine triphosphate and adenosine in the mammalian heart: Clinical and experimental aspects. Journal of the American College of Cardiology, 4, 414.

Belhassen, B., Pelleg, A., Shoshani, D. & Laniado, S., 1984. Atrial fibrillation induced by adenosine triphosphate. American Journal of Cardiology, 53, 1405.

Belhassen, B. & Pelleg, A., 1985. Adenosine triphosphate and adenosine: Perspectives in the acute management of paroxysmal supraventricular tachycardia. Clinical Cardiology, 8, 460.

Brown, J.H. & Jones, L.G., 1986. Phosphoinositide metabolism in the heart. In Phosphoinositides and Receptor Mechanisms, edited by J.W. Paton, Jr., (Alan R. Liss, Inc., New York), p. 245.

Burnstock, G., 1980. Purinergic receptors in the heart. Circulation Research, 46, 176.

Chiba, S., Kubata, K. & Hashimoto, K., 1973. European Journal of Pharmacology, 21, 281.

Clemo, H.F. & Belardinelli, L., 1986a. Effect of adenosine on atrioventricular conduction. I. Site and characterization of adenosine action in the guinea pig atrioventricular node. Circulation Research, 59, 427.

Clemo, H.F. & Belardinelli, L., 1986b. Effect of adenosine on atrioventricular conduction. II. Modulation of atrioventricular node transmission by adenosine in hypoxic isolated guinea pig hearts. Circulation Research, 59, 437.

Collis, M. & Pettinger, S.J., 1982. Can ATP stimulate P_1-receptors in guinea pig atrium without conversion to adenosine. European Journal of Pharmacology, 81, 521.

Cuparencu, B., Safts, L., Sandor, V.I., Fabian, A., Horak, J. & Marmo, E., 1986. Central antiarrhythmic activity of adenosine in anesthetized rats. Research Communications in Chemical Pathology and Pharmacology, 51, 47.

DiMarco, J.P., Sellers, D., Berne, R.M., West, G.A. & Belardinelli, L., 1983. Adenosine: Electrophysiologic effects and therapeutic use for terminating paroxysmal supraventricular tachycardia. Circulation, 68, 1254.

DiMarco, J.P., 1987. Adenosine and supraventricular tachycardia. In Cardiac Electrophysiology and Pharmacology of Adenosine and ATP: Basic and Clinical Aspects, edited by A. Pelleg, E.L. Michelson, & L.S. Dreifus, (Alan R. Liss, Inc., New York), p. 271.

Dobson Jr., J.G., Fenton, R.A. & Romano, F.D., 1987. The

antiadrenergic actions of adenosine in the heart. In Topics and Perspectives in Adenosine Research, edited by E. Gerlach & B.F. Becker (Springer-Verlag, Berlin), p. 356.

Drury, A.N. & Szent-Gyorgyi, A., 1929. The physiological activity of adenine compounds with especial reference to their action upon the mammalian heart. Journal of Physiology (London), 68, 213.

Fagbemi, O. & Parratt J.R., 1984. Antiarrhythmic action of adenosine in the early stages of experimental myocardial ischemia. European Journal of Pharmacology, 100, 243.

Ford, D.A., Sharp, J.A. & Rovetto, M.J., 1985. Erythrocyte adenosine transport: Effects of Ca^{2+} channel antagonists and ions. American Journal of Physiology, 248, H593.

Fujita, S., Ishida, Y., Izumi, K., Moritoki, H., Ohara, M. & Takei, M., 1980. Potentiation by dilazep of the negative inotropic effects of adenosine on guinea-pig atria. British Journal of Pharmacology, 68, 343.

Gilmour, R.F. & Zipes, D.P., 1985. Positive inotropic effect of acetylcholine in canine cardiac Purkinje fibers. American Journal of Physiology, 249, H735.

Hartzell, H.C., 1979. Adenosine receptors in frog sinus venosus: Slow inhibitory potentials reproduced by adenosine compounds and acetylcholine. Journal of Physiology (London), 293, 23.

Hopwood, A.M., Harding, S.E. & Harris, P., 1987. An antiadrenergic effect of adenosine on guinea-pig but not rabbit ventricles. European Journal of Pharmacology, 137, 67.

Huang, T.F. & Yang, B.B., 1985. Effect of vagotomy and drugs on the ventricular fibrillation threshold of normal and ischemic hearts in the anesthetized rat. Archives Internationales de Pharmacodynamie et de Therapie, 278, 128.

James, T.N., 1965. The chronotropic action of ATP and related compounds studied by direct perfusion of the sinus node. Journal of Pharmacology and Experimental Therapeutics, 149, 233.

Jochem, G. & Nawrath, H., 1983. Adenosine activates potassium conductance in guinea pig atrial heart muscle. Experientia, 39, 1347.

Kolassa, N., Pfleger, K. & Rummel, W., 1970. Specificity of adenosine uptake into the heart and inhibition by dipyridamole. European Journal of Pharmacology, 9, 265.

Kurachi, Y., Nakajima, T. & Sugimoto, T., 1986. On the mechanism of activation of muscarinic K^+ channels by adenosine in isolated atrial cells: Involvement of GTP-binding protein. Pflugers Archiv, 407, 267.

Laorden, M.D., Hernandez, J. & Ribeiro, J.A., 1986. The effects of adenosine, ATP and ADP on ventricular automaticity induced by a local injury in the isolated right ventricle of the rat. Archives Internationales de Pharmacodynamie et de Therapie, 279, 258.

Lerman, B.B., Belardinelli, L., West, G.A., Berne, R.M. & DiMarco, J.P., 1986. Adenosine-sensitive ventricular tachycardia: Evidence suggesting cyclic AMP-mediated triggered activity. Circulation, 74, 270.

Luckhoff, A. & Busse, R., 1986. Increased free calcium in endothelial cells under stimulation with adenine nucleotides. Journal of Cellular Physiology, 126, 414.

Marangos, P.J., Finkel, S.M., Verma, A., Maturi, M.F., Patel, J. & Patterson, R.E., 1984. Adenosine uptake sites in dog heart and brain interaction with calcium antagonists. Life Sciences, 35, 1109.

Martin, J.M., Subers, E.M., Halvorsen, S.W. & Nathanson, N.M., 1986. Functional and physical properties of chick atrial and ventricular GTP-bindings proteins: Relationship to muscarinic acetylcholine receptor-mediated responses. Journal of Pharmacology and Experimental Therapeutics, 240, 683.

Masse, C., Sassine, A., Koliopoulos, N. & Munoz, A., 1986. Positive inotropic effect of adenosine in rabbit atria. Pflugers Archiv, 406, 551.

Masters, S.B., Quin, M.T. & Brown, J.H., 1985. Agonist-induced desensitization of muscarinic receptor-mediated calcium efflux without concomitant desensitization of phosphoinositide hydrolysis. Molecular Pharmacology, 27, 325.

Mazgalev, T., Dreifus, L.S., Michelson, E.L. & Pelleg, A., 1987. Adenosine-vagus interaction at the cellular level. In Cardiac Electrophysiology and Pharmacology of Adenosine and ATP: Basic and Clinical Aspects, edited by A. Pelleg, E.L. Michelson & L.S. Dreifus (Alan R. Liss, Inc., New York), p. 195.

Meszaros, J., Kelemen, K., Kecskemeti, V. & Szegi, J., 1984. Antagonism between adenosine and brombobenzoyl-methyladamantylamine, a K^+ channel blocker, in atrial myocardium of guinea pig. European Journal of Pharmacology, 98, 265.

Meszaros, J., Kelemen, K., Kecskemeti, V. & Kovacs, G., 1986. Interaction between adenosine and antiarrhythmic agents in atrial myocardium of guinea-pig. Archives Internationales de Pharmacodynamie et de Therapie, 280, 84.

Motomura, S. & Hashimoto, K., 1982. Reperfusion-induced bradycardia in the isolated, blood-perfused sino-atrial node and papillary muscle preparation of the dog. Japanese Heart Journal, 23, 112.

Munoz, A., Sassine, A., Lehujeur, C., Koliopoulos, N. & Puech, P., 1985. Mode d'action des substances purinergiques derivees de l'adenine sur la conduction auriculo-ventriculaire. Archives de Malaise Coeur, 78, 143.

Munoz, A., Sassine, A. & Puech, P., 1987. Cardiac electropharmacology and therapeutic use of adenosine and adenosine triphosphate. In Cardiac Electrophysiology and Pharmacology of Adenosine and ATP: Basic and Clinical Aspects, edited by A. Pelleg, E.L. Michelson & L.S. Dreifus (Alan R. Liss, Inc., New York), p. 255.

Needleman, P., Minkes, M.S. & Douglas Jr., J.R., 1974. Stimulation of prostaglandin biosynthesis of adenine nucleotides. Profile of prostaglandin release by perfused organs. Circulation Research, 34, 455.

Paton, D.M. & Kurahashi, K., 1981. Structure-activity relations for negative chronotropic action of adenosine in isolated rat atria: Evidence for an action on A_1 receptors. IRCS Medical Science, 9, 447.

Pelleg, A., 1985. Cardiac cellular electrophysiologic actions of adenosine and adenosine triphosphate. American Heart Journal, 110, 688.

Pelleg, A., 1987. Adenosine and ATP interactions with the autonomic nervous system: Modulation of their cardiac electrophysiology and pharmacology. Clinical Pharmacology, 27, 366.

Pelleg, A., Belhassen, B., Ilia, R. & Laniado, S., 1985a. Comparative electrophysiologic effects of ATP and adenosine in the canine heart: Influence of atropine, propranolol, vagotomy, dipyridamole and aminophylline. American Journal of Cardiology, 55, 571.

Pelleg, A., Mitamura, H. & Michelson, E.L., 1985b. Evidence for vagal involvement in the electrophysiologic actions of exogenous adenosine and adenosine triphosphate in the canine heart. Journal of Autonomic Pharmacology, 5, 207.

Pelleg, A., Mitamura, H., Michelson, E.L. & Dreifus, L.S., 1986a. Evidence against prostaglandin mediation of the electrophysiologic effects of ATP versus adenosine in the canine heart. Journal of Cardiovascular Pharmacology, 8, 534.

Pelleg, A., Mitamura, H., Mitsuoka, T., Michelson, E.L. & Dreifus, L.S., 1986b. Adenosine and ATP suppress ventricular escape rhythms in the canine heart. Journal of the American College of Cardiology, 8, 1145.

Pelleg, A., Mitsuoka, T. & Michelson, E.L., 1987a. Adenosine mediates the negative chronotropic action of adenosine 5'-triphosphate in the canine sinus node. Journal of Pharmacology and Experimental Therapeutics, 242, 791.

Pelleg, A. & Michelson, E.L., 1987b. Vagally mediated Ca^{2+} effects on the depressant actions of adenosine and adenosine 5'-triphosphate on the canine sinus node in vivo. Journal of Autonomic Pharmacology, 7, 1.

Pelleg, A., Mitsuoka, T., Mazgalev, T., Michelson, E.L. & Dreifus, L.S., 1987c. Integrative negative chronotropic effects of adenosine and the vagus nerve on the canine sinus node. Cardiovascular Research (in the press).

Pelleg, A., Mitamura, H., Mitsuoka, T., Mazgalev, T., Michelson, E.L. & Dreifus, L.S., 1987d. Interactive negative chronotropic actions of adenosine and verapamil on the canine sinus node in vivo. In Cardiac Electrophysiology and Pharmacology of Adenosine and ATP: Basic and Clinical Aspects, edited by A. Pelleg, E.L. Michelson, & L.S. Dreifus (Alan R. Liss, Inc., New York), p. 235.

Phillis, J.W. & Wu, P.H., 1981. Indomethacin, ibuprofen, and meclofenamate inhibit adenosine uptake by rat brain synaptosomes. European Journal of Pharmacology, 72, 139.

Quist, E.E. & Satumtira, N., 1987. Muscarinic receptor stimulated phosphoinositide turnover in cardiac atrial tissue. Biochemical Pharmacology, 36, 499.

Raberger, G., Fischer, G., Krumpl, G., Schneider, W. & Stroisnig, H., 1987. Further evidence against adenosine-catecholamine antagonism in vivo. Investigations with treadmill exercise in dogs. In Topics and Perspectives in Adenosine Research, edited by E. Gerlach & B.F. Becker (Springer-Verlag, Berlin), p. 383.

Rand, M., Stafford, A. & Thorp R.H., 1955. The potentiation of the action of adenosine on the guinea pig heart by ouabain. Journal of Pharmacology and Experimental Therapeutics, 114, 119.

Rotmensch, H.H., Bianchi, P.C., Narayan, S., Grau, T. & Ferguson, R.K., 1982. Verapamil inhibition of adenosine receptor-dependent calcium uptake. Circulation, 66, II-140.

Rubio, R., Bianchi J., Berne, R.M., 1986. Atrial negative inotropic effect of acetylcholine (Ach) and adenosine (ADO) are the result of the initial trigger of a positive inotropic mechanism. Pflugers Archiv, 407, S53.

Schrader, J., Rubio, R. & Berne, R.M., 1975. Inhibition of slow action potentials of guinea pig muscle by adenosine: A possible effect on Ca^{+} influx. Journal of Molecular and Cellular Cardiology, 7, 427.

Seitelberger, R., Schutz, W., Schlappack, O. & Raberger, G.,

1984. Evidence against the adenosine catecholamine antagonism under in vivo conditions. Archiv Pharmacologie, 325, 234.

Shah, P., Nalos, P. & Peter, T., 1987. Atropine resistant post infarction complete AV block: Possible role of adenosine and improvement with aminophylline. American Heart Journal, 113, 194.

Sohn, Y.Z., Hong, J.C. & Katz, R.L., 1982. Antiarrhythmic and hemodynamic responses to adenosine triphosphate during infusion of epinephrine in dogs anesthetized with halothane. Anesthesia and Analgesia, 61, 423.

Solti, F., Juhasz-Nagy, S., Kecskemeti, V. & Czako, E., 1984. The effect of adenosine on impulse formation and propagation in the heart. Cor Vasa, 26, 296.

Stafford, A., 1966. Potentiation of adenosine and adenine nucleotides by dipyridamole. British Journal of Pharmacology and Chemotherapy, 28, 218.

Szentmiklosi, A.J., Nemeth, M., Szegi, J., Papp, J.G. & Szekeres, L., 1979. On the possible role of adenosine in hypoxia-induced alterations of the electrical and mechanical activity of the atrial myocardium. Archives Internationales de Pharmacodynamie et de Therapie, 238, 283.

Urthaler, F., Woods, W.T., James, T.N. & Walker, A.A., 1981. Effects of adenosine on mechanical performance and electrical activity in the canine heart. Journal of Pharmacology and Experimental Therapeutics, 216, 254.

Van Belle, H., 1985. Myocardial purines during ischemia, reperfusion and pharmacological protection. Molecular Physiology, 8, 615.

Watanabe, I., Saito, T., Saito, S., Tokutake, E., Hibiya, K., Ozawa, Y. & Hatano, M., 1983. Role of adenosine on atrioventricular conduction disturbances during myocardial ischemia. Pace, 6, A-7.

Wayne, E.J., Goodwin, J.F. & Stoner, H.B., 1949. The effect of adenosine triphosphate on the electrocardiogram of man and animals. British Heart Journal, 11, 55.

Wesley Jr., R.C., Boykin, M.T. & Belardinelli, L., 1986. Role of adenosine as mediator of bradyarrhythmias during hypoxia in isolated guinea pig hearts. Cardiovascular Research, 20, 752.

West, G.A. & Belardinelli, L., 1985. Correlation of sinus slowing and hyperpolarization caused by adenosine in sinus node. Pflugers Archiv, 403, 75.

Zehl, U., Ritter, C. & Forster, W., 1976. Influence of prostaglandins upon adenosine release and of adenosine upon prostaglandin release in the isolated rabbit heart. Acta Biologica Germanica, 35, K77.

IS ADENOSINE AN ENDOGENOUS ANTIARRHYTHMIC AGENT UNDER CONDITIONS OF MYOCARDIAL ISCHAEMIA?

James R Parratt, Gabriel Boachie-Ansah,
Kathleen A Kane and Cherry L Wainwright

The Cardiovascular Unit, Department of Physiology & Pharmacology, University of Strathclyde, Glasgow G1 1XW, Scotland, UK

INTRODUCTION

A number of highly vasoactive agents are released under conditions of myocardial ischaemia which might modify the genesis, course and severity of early ischaemia-induced ventricular arrhythmias. Among these substances are some that are almost certainly proarrhythmic. These include noradrenaline (Riemersma and Dart, 1985), thromboxane (Parratt et al. 1985), and possibly, 5-hydroxytryptamine (Coker et al. 1986) and opioid peptides (Parratt & Sitsapesan 1986). However, it is also possible that during ischaemia the heart liberates 'protective' endogenous substances which might reduce the severity of these arrhythmias. This concept of endogenous antiarrhythmic agents was first suggested by Werner Förster in 1976 (Förster 1976).

A number of conditions would need to be fulfilled before a substance could be considered as an endogenous antiarrhythmic agent: (i) it would need to be formed and released in appropriate concentrations during ischaemia, (ii) its exogenous (perhaps local) administration should reduce the severity of ischaemia-induced arrhythmias, (iii) agents that increase its formation, inhibit its breakdown or otherwise increase its concentration around the relevant receptors should have antiarrhythmic properties, and (iv) drugs that block the receptor upon which such endogenous agents act should intensify the severity of these arrhythmias.

We have provided evidence that prostacyclin should be considered as such an endogenous antiarrhythmic substance and have suggested that one rational approach to antiarrhythmic therapy would be to attempt to 'boost' the amounts of this arachidonic acid metabolite produced and released under conditions of myocardial ischaemia. There are a number of such prostacyclin promoters (Parratt 1987) including nafazatrom (Coker & Parratt 1984). Adenosine is another possible candidate as an endogenous antiarrhythmic agent and indeed we have previously suggested that 'released adenosine might play a protective role in the early stages of myocardial ischaemia and might act as an endogenous antiarrhythmic' (Fagbemi & Parratt 1984). In summary the evidence is:

i. Adenosine exerts a profound antiarrhythmic effect in at least two species when infused either intravenously or by way of the lumen of the left ventricle, under conditions of myocardial ischaemia and reperfusion (Fagbemi & Parratt 1984; Wainwright & Parratt 1987; and present results).

ii. It is locally released (coronary venous sampling from the ischaemic region following coronary artery occlusion in concentrations that are similar to those obtained when adenosine is infused into the left ventricle in amounts sufficient to exert an anti-arrhythmic effect (Wainwright & Parratt 1987).

iii. Potentiation of the biological effects of endogenously released adenosine with dipyridamole is an anti-arrhythmic procedure (Wainwright & Parratt 1987).

METHODS

(1) Studies in anaesthetised rats

Male Sprague Dawley rats (250-350g) were anaesthetised with sodium pentobarbitone, artificially ventilated and thoracotomised. The left coronary artery was ligated, and the early ischaemic arrhythmias were assessed, as described in detail by Clark et al. (1980) and by Fagbemi & Parratt (1984). In those studies in which compounds were infused locally a catheter was also placed in the lumen of the left ventricle by way of the right common carotid artery to allow for direct left ventricular drug administration (ilv).

Adenosine was infused either intravenously (iv), in doses ranging from 2.5μg $kg^{-1}min^{-1}$, or directly into the left ventricle (ilv, 5 and 10μg $kg^{-1}min^{-1}$) commencing 15 minutes prior to coronary artery occlusion and maintained to the end of the 30 minute occlusion period. Similar studies were performed with ATP (sodium salt, vanadium free) in doses of 10, 25 and 50μg $kg^{-1}min^{-1}$, ADP (sodium salt) in doses of 10 and 50μg $kg^{-1}min^{-1}$, AMP (10, 50, 250μg $kg^{-1}min^{-1}$), inosine (10, 50, 250 μg $kg^{-1}min^{-1}$) and hypoxanthine (10 and 50μg $kg^{-1}min^{-1}$).

(2) Studies in anaesthetised dogs

Greyhounds of either sex were anaesthetised with sodium thiopentone followed by chloralose and prepared for coronary artery occlusion and haemodynamic measurements as described in detail by Coker and Parratt (1984) and by Wainwright & Parratt (1987).

After preliminary studies, a dose of adenosine of 10μg $kg^{-1}min^{-1}$ was chosen and infused into the left ventricle for a total period of 50 minutes, commencing 15 minutes prior to coronary artery occlusion. This dose resulted in a sustained increase in coronary blood flow. Studies were also performed with dipyridamole in a dose (0.25μg kg^{-1} iv) which caused a 10-fold shift to the left of adenosine dose-response curves.

(3) Electrophysiological studies in sheep Purkinje fibres

These preparations were superfused with warmed Tyrode solution, gassed with 95% O_2 and 5% CO_2 and stimulated at 1.5 Hertz through bipolar silver electrodes. Glass microelectrodes filled with 3M KCl were used to measure transmembrane potentials which were stored on film and subsequently analysed to evaluate resting membrane potential (RMP), maximum rate of depolarisation of phase 0 (MRD), action potential amplitude (APA), overshoot (OS) and duration (APD). Studies were also performed under conditions designed to simulate 'mild' ischaemia by increasing the KCl from 5.4 to 8.0mM, lowering the pH from 7.4 to 6.8 units and gassing with 95% N_2 and 5% CO_2 (Boachie-Ansah et al. 1986).

RESULTS

(1) Studies in anaesthetised rats

Adenosine, given either as an intravenous infusion or directly into the left ventricle, significantly reduced the severity of ischaemia-induced arrhythmias. When adenosine was given intravenously, there were dose-dependent reductions in the number of ventricular ectopic beats and in the duration of VT (Fagbemi & Parratt 1984) and these effects were even more pronounced when adenosine was infused directly into the lumen of the left ventricle (Table 1) although the incidence of VF and mortality were only reduced by the lowest dose used (5μg $kg^{-1}min^{-1}$). The reduction in ectopic activity occurred throughout the entire 0-30 min occlusion period but was especially pronounced during the 10-15 min occlusion period when arrhythmias are especially severe. When administered in this way adenosine only transiently reduced arterial pressure; with continued infusion the arterial pressure was not significantly different from controls either immediately preceding or throughout the period of coronary artery occlusion. Heart rate was lower than in the controls but not significantly so.

Neither ADP, AMP nor inosine were effective in this model when infused directly into the lumen of the left ventricle but some protection was afforded by both ATP and hypoxanthine. ATP was effective in doses (25 and 50μg $kg^{-1}min^{-1}$; Table 1), which had no significant effect on either blood pressure or heart rate. Hypoxanthine exerted an antiarrhythmic effect in doses of 10 and 50μg $kg^{-1}min^{-1}$ (Table 1); again, there were no significant haemodynamic effects.

Attempts were made to inhibit cellular adenosine transport with dilazep (10 and 25μg kg^{-1}). This treatment tended to increase all the indices of arrhythmia severity.

(2) Studies in anaesthetised greyhound dogs

Ventricular arrhythmias after coronary artery occlusion in this species include single and multiple extrasystoles, ventricular tachycardia (VT) and occasionally, ventricular fibrillation (VF). They commence 1-2 min after occlusion and there are two distinct periods, 3-7 min post-occlusion (phase 1a) and 12-17 min post-occlusion (phase 1b), when ventricular ectopic activity is particularly pronounced. The left ventricular infusion of adenosine (10μg $kg^{-1}min^{-1}$) significantly reduced the total number of ectopic beats and was especially effective against salvos and ventricular tachycardia. This is illustrated in Figure 1. Ventricular ectopic beats (VEB's) as salvos were reduced from 194±40 to only 27±15 ($P < 0.01$) and as VT from 379±112 to 74±36 ($P < 0.05$). Total VEB's were reduced from 786±115 in the controls to 156±44 in the treated group ($P < 0.001$). This dose of adenosine significantly reduced mean arterial blood pressure (from 130±12 to 113±14 mmHg) without modifying heart rate, pulmonary artery pressure, LVEDP, LVdP/dT max or cardiac output.

Dipyridamole (0.25mg/kg) also exerted significant anti-arrhythmic effects. VEB's as salvos were reduced to 66±38 ($P < 0.05$) and total ectopic beats to 388±167 ($P < 0.05$). The incidence of VT was reduced by both adenosine and dipyridamole (from 82% to 43% and 44% respectively). The dose of dipyridamole used had significant haemodynamic effects. Blood pressure was reduced from 163±13 to 142±13 mmHg ($P < 0.01$) and LVEDP was reduced (from 7.3±1.3

Table 1. Effects of adenosine, ATP and hypoxanthine (by left ventricular infusion) on the incidence and severity of coronary artery occlusion-induced ventricular arrhythmias in anaesthetised rats.

	Ventricular ectopic beats	Ventricular tachycardia(VT)		Ventricular fibrillation(VF)	
		duration(s)	incidence (%)	incidence (%)	mortality (%)
Control	1515 ± 358	94 ± 27	100	90	30
Adenosine					
5µg kg^{-1} min^{-1}	829 ± 210*	59 ± 24	90	50	0
10µg kg^{-1} min^{-1}	385 ± 155**	40 ± 16	80	50	40
Control	908 ± 206	55 ± 14	91	64	10
ATP					
10µg kg^{-1} min^{-1}	448 ± 126	23 ± 9	80	30	10
25µg kg^{-1} min^{-1}	242 ± 61*	7 ± 2*	80	40	20
50µg kg^{-1} min^{-1}	380 ± 152*	29 ± 15	70	30	0
Control	1585 ± 450	106 ± 34	100	80	30
Hypoxanthine					
10µg kg^{-1} min^{-1}	822 ± 310*	58 ± 26	90	60	20
50µg kg^{-1} min^{-1}	250 ± 81**	8 ± 3**	77	33*	20

n = 9-10 for each group; * $P < 0.05$; ** $P < 0.01$. Values are mean ± s.e.m.

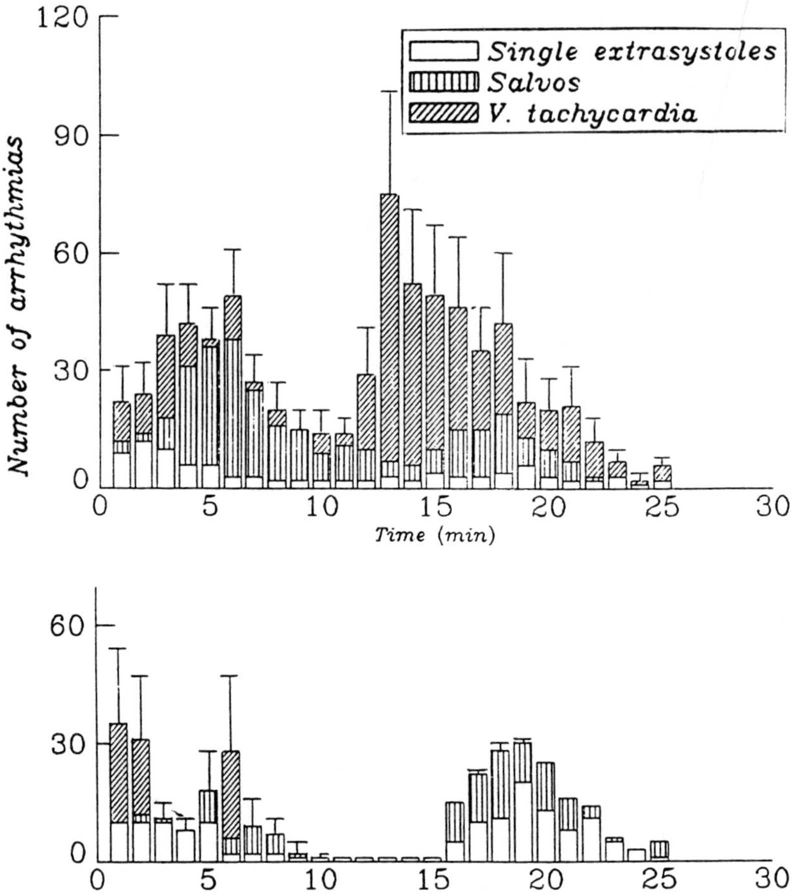

Figure 1. The distribution of different types of ventricular arrhythmias following coronary artery occlusion in anaesthetised greyhound dogs (above) and in dogs receiving adenosine (10 mgKg^{-1}min^{-1} ilv; below). In control dogs there are two distinct phases of particularly severe ectopic activity, phase 1a (between about 3 and 6 min of the commencement of the occlusion) and phase 1b (between about 12 and 18 min). VT is especially prevalent during phase 1b. Adenosine markedly reduced ectopic activity throughout the ischaemic period with a particular effect on VT during phase 1b.

to 5.5±1.2 mmHg $P < 0.01$). There was an increase in cardiac output from 2.65± 0.18 to 3.00±0.17 l/min ($P < 0.05$).

(3) <u>Electrophysiological studies in sheep Purkinje fibres</u>
In normal fibres adenosine caused no significant changes in RMP, MRD, APA or OS. There was a slight, concentration-dependent, shortening of action potential duration at both 50% and 90% repolarisation with adenosine concentrations of 10^{-6} - 10^{-4}M. ATP had similar, though less pronounced, effects and RMP was also significantly reduced with the highest concentrations used (10^{-4}M). More pronounced effects were observed in 'mildly ischaemic' fibres. The ischaemic solution itself induced marked changes in all the measured parameters (Figure 2) with reductions in RMP, MRD, APA, OS and APD. Some of these changes (e.g. the reductions in MRD and APA) were significantly attenuated in the presence of adenosine or ATP (10^{-5} - 10^{-4}M).

DISCUSSION

The results of this study indicate that adenosine, when infused in appropriate concentrations into the lumen of the left ventricle, reduces the severity of <u>early</u> ischaemic ventricular arrhythmias in two different species. This confirms our earlier studies using anaesthetised rats subjected to coronary artery ligation in which adenosine was infused intravenously (Fagbemi & Parratt 1984). In addition, in one of these species, ATP and hypoxanthine were also effective, although rather less so than adenosine itself. There is now growing evidence demonstrating the beneficial effects of adenosine against abnormal ventricular rhythms. These include ischaemia-induced arrhythmias in rat isolated hearts (Lubbe et al. 1983), where protection was attributed to a reduction in ischaemia-induced increases in myocardial cyclic AMP, adrenaline-induced arrhythmias in dogs anaesthetised with halothane (Sohn et al. 1982) and ventricular escape rhythms in dog hearts following electro-ablation of the His bundle region (Pelleg et al. 1986).

There are many possible mechanisms for these protective effects. These include improved perfusion to the ischaemic areas or, more likely, to regions bordering the area of pronounced blood flow reduction, inhibition of the release and/or effects at receptor level of catecholamines, increased prostacyclin production (Ciabattoni and Wennmalm 1985) indicating that adenosine should be added to the increasing list of 'prostacyclin promoters', and a reduction in cardiac work. Adenosine also has pronounced efects on cardiac muscle action potentials (reviewed by Pelleg 1985 and by Isenberg et al. 1987) that could explain, or contribute to, an antiarrhythmic effect in vivo. These

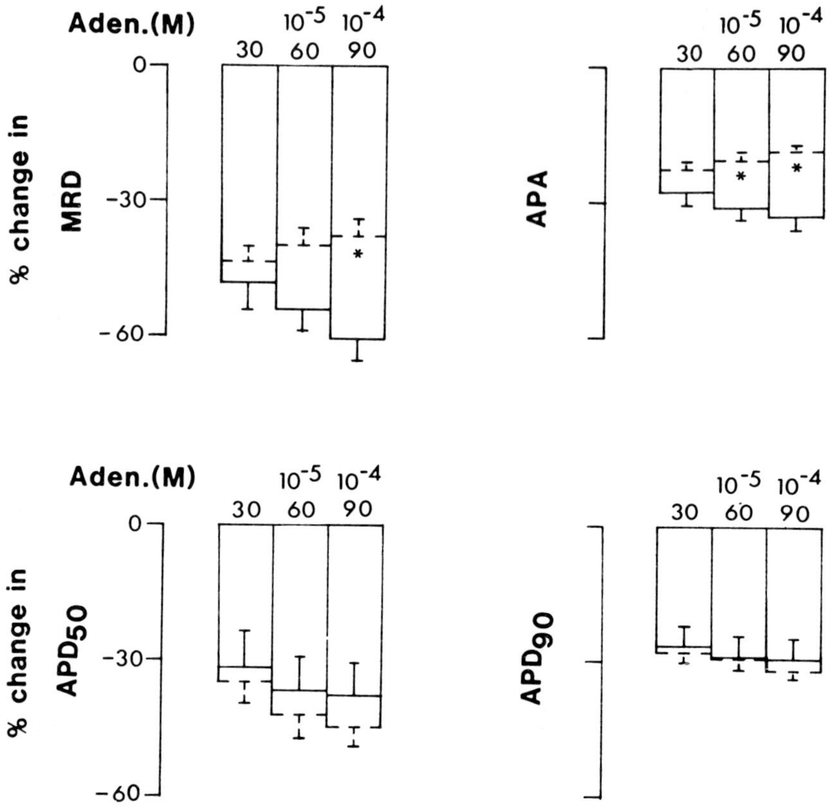

Figure 2. Effects of adenosine (10^{-5} and 10^{-4}M; ...) on the changes in transmembrane action potentials recorded from sheep Purkinje fibres subjected to simulated mild ischaemia (——). Ischaemia alone resulted in time-dependent reductions in the maximum rate of depolarisation of phase 0 (MRD), action potential amplitude (APA), action potential duration (APD_{50} and APD_{90} and in the resting membrane potential (RMP). Some of the changes (e.g. MRD, APA were attenuated by adenosine, especially following prolonged 'ischaemia'.

actions include suppression of automaticity (Szentmiklosi et al. 1980; Rosen et al. 1983), a reduction in MRD (without modifying the resting membrane potential; Urthaler et al. 1981) and antagonism of the catecholamine-stimulated Ca^{++} inward current (I_{Ca}; Isenberg et al. 1987).
The studies described in the present paper indicate that, whereas adenosine does not significantly modify electrophysiological parameters in normal Purkinje fibres (except for a slight reduction in APD), under conditions designed to mimic mild ischaemia both adenosine and ATP exert electrophysiological effects which, if they were also evident in vivo, could partly explain an antiarrhythmic effect during ischaemia. These effects include an attenuation of the reduction in MRD and of the shortening of the action potential which occur during ischaemia. It is noteworthy that whereas adenosine does not directly alter membrane currents in ventricular myocytes, it does attenuate alterations in Ca^{++} influx resulting from β-adrenoceptor stimulation (Isenberg et al. 1987). This effect is apparently mediated by A_1 receptors coupled to adenylate cyclase via a GTP-binding protein. Whether noradrenaline release occurs in the preparation we have used and contributes to the electro- physiological effects of mild ischaemia is not known.

We have not, of course, answered the question that forms the title to this paper. It is possible that adenosine, released under conditions of blood flow reduction, is a protective factor elaborated by the myocardium and that it is concerned not only with the tight relationship between myocardial oxygen demand and coronary blood flow (reviewed by Berne 1980; Parratt 1982) but also with reducing the severity of ischaemia-induced ventricular arrhythmias. The only direct, albeit inconclusive, evidence we have for this concept is the antiarrhythmic effect of dipyridamole in the dog model. Other approaches under consideration or active investigation include the use of adenosine receptor blocking drugs (which should intensify ischaemia-induced arrhythmias) and altering nucleoside transport through membranes.

These results may have clinical relevance. "Thus, it could turn out that the naturally occurring biologic compounds adenosine and ATP can be used as antiarrhythmic drugs in the ventricles, in addition to exerting their recognised antiarrhythmic effects on paroxysmal supraventricular tachycardia incorporating the AV node" (Pelleg et al. 1986). It is of interest in this regard that ventricular tachycardia was the arrhythmia most susceptible to adenosine in our dog studies as well as in the one clinical report demonstrating a beneficial effect of adenosine against ventricular arrhythmias (Lerman et al. 1986).

The work reported in this paper was variously supported by the Scottish Home & Health Department.

REFERENCES

Berne, R.M., 1980, Circulation Research, 47, 807.

Boachie-Ansah, G., Kane, K.A. & Parratt, J.R., British Journal of Pharmacology, 88, 309P.

Ciabottoni R. & Wennmalm, A., 1985, British Journal of Pharmacology, 85, 557.

Clark, C. et al., 1980, Journal of Pharmacological Methods, 76, 504.

Coker, S.J. & Parratt, J.R., 1984, Journal of Molecular and Cellular Cardiology, 16, 43.

Coker, S.J. et al., 1986. European Journal of Pharmacology, 127, 211.

Fagbemi, O. & Parratt, J.R., 1984, European Journal of Pharmacology, 100, 243.

Förster, W., 1976, Acta biologica et medica germanica 35, 1101.

Isenberg, G. et al., 1987. In Topics and Perspective in Adenosine Research, edited by E. Gerlach & B.F. Becker (Springer, Berlin), p. 323.

Lerman, B.B. et al., 1986, Circulation, 74, 270.

Lubbe, W.F. et al., 1983, Federation Proceedings, 42, 2460.

Parratt, J.R., 1982. In The Coronary Artery, edited by S. Kalsner (Croom Helm, London), p. 309.

Parratt, J.R., 1987, Handbook of Experimental Pharmacology, edited by E.M. Vaughan Williams, 89, (in press).

Parratt, J.R. & Coker, S.J., 1985, Journal of Cardiovascular Pharmacology, 7, Suppl. 5, S65.

Parratt, J.R. & Sitsapesan, R., 1986, British Journal of Pharmacology, 87, 621.

Pelleg, A., 1985, American Heart Journal, 110, 688.

Pelleg, A. et al., 1986, Journal of the American College of Cardiology, 8, 1145.

Riemersma, R.A. & Dart, A.M. (editors), 1985, Journal of Cardiovascular Pharmacology, 7, Suppl. 5.

Rosen, M.R. et al., 1983, American Journal of Physiology 244, H1715.

Sohn, Y.Z. et al., 1982, Anaesthesia and Analgesia, 61, 423.

Szentmiklosi, A.J. et al., 1980, Archives Internationale Pharmacodynamie, 311, 147.

Urthaler, F. et al., 1981, Journal of Pharmacology and Experimental Therapetuics, 216, 254.

Wainwright, C.L. & Parratt, J.R., 1987, European Journal of Pharmacology, (in press).

ON THE MECHANISM OF ACTION OF ADENOSINE ON MYOCARDIAL CONTRACTILITY

Wilhelm Schmitz, Thomas Heller, Martina Köcher, Joachim Neumann, Hasso Scholz, Jens Scholz, Volker Stemmildt & Katharina Störtzel

Abteilung Allgemeine Pharmakologie, Universitäts-Krankenhaus Eppendorf, Martinistraße 52, D-2000 Hamburg 20, FRG

INTRODUCTION

Adenosine has quite different effects on myocardial contractility in different parts of the heart (Drury & Szent-Györgyi 1929). In atrial preparations adenosine and adenosine receptor agonists have a pronounced negative inotropic effect whereas they have no or a slight positive inotropic effect in ventricular preparations (Belardinelli & Isenberg 1983a, 1983b; Böhm et al. 1984; Brückner et al. 1985; Nawrath et al. 1985). However, in ventricular heart muscle preparations adenosine and the adenosine analogs (-)-N^6-phenylisopropyladenosine (PIA) and 5'-N-ethylcarboxamideadenosine (NECA) antagonize the positive inotropic effect of substances which increase cardiac cAMP content, i.e. they are negatively inotropic under these conditions (Schrader et al. 1977; Dobson 1978; Baumann et al. 1981; Böhm et al. 1984, 1985; Schmitz et al. 1985). The negative inotropic effects of adenosine and its analogs in atrial preparations which occur without treatment with cAMP-increasing agents are termed "direct" while the negative inotropic effects of the nucleosides in ventricular preparations which are detectable in the presence of cAMP-increasing agents only are designated "indirect" (Endoh et al. 1983a). The mechanism of the "direct" negative inotropic effect of adenosine is not related to changes in cAMP (Endoh et al. 1983a, 1983b; Böhm et al. 1984; Brückner et al. 1985). Concerning the mechanism of the "indirect" negative inotropic effects of adenosine and its analogs it is still a matter of debate whether or not it is due to an attenuation of the cAMP-increase by adenylate cyclase stimulators or PDE inhibitors (Schrader et al. 1977; Dobson 1983; Böhm et al. 1984, 1985; Schmitz et al. 1985; Linden et al. 1985). A decrease of the isoprenaline-induced increase in cAMP (Schrader et al. 1977; Dobson 1983), no change in cAMP (Böhm et al. 1984; Schmitz et al. 1985) or even an increase in cAMP (Huang & Drummond 1976; Anand-Srivastava 1985) have been reported.

We investigated the effects of adenosine and PIA on atrial force of contraction and action potential. The effects of adenosine on phosphoinositide turnover were also studied. In ventricular preparations the effects of PIA and NECA in the presence of the beta-adrenoceptor agonist isoprenaline on force of contraction, on adenylate cyclase activity and on cAMP content in intact contracting papillary muscles were investigated. Furthermore, the effects of the nucleosides on slow action potentials (as a measure of slow Ca^{++}-inward current) were studied. Finally, the influence of pertussis toxin treatment was studied in order to investigate whether a nucleotide-binding protein (N-protein) is involved in the signal transduction of the adenosine receptor mediated effects.

METHODS

The contraction experiments were performed on isolated electrically driven (1 Hz) left auricles or papillary muscles from guinea-pig hearts under isometric conditions as described by Böhm et al. (1984, 1985). Normal transmembrane action potentials and slow action potentials were recorded with a conventional microelectrode technique (Brückner et al. 1980; Böhm et al. 1985). cAMP content was measured by radioimmunoassay (Böhm et al. 1984). Adenylate cyclase activity was determined in washed particulate fractions (10,000 x g sediment) from homogenates of right ventricles from guinea-pig hearts (Böhm et al. 1984). Phosphatidylinositol 4,5-bisphosphate (PIP_2) and inositol 1,4,5-trisphosphate (IP_3) were determined with the methods of Berridge et al. (1982) and Creba et al. (1983) in electrically driven (1 Hz) left auricles from rats as described previously by Schmitz et al. (1987a, 1987b).

RESULTS

Effects of adenosine and PIA on atrial preparations

The direct negative inotropic effect of PIA is accompanied by a marked shortening of the action potential duration (fig. 1 A). The negative inotropic and the action potential shortening effects are abolished by pertussis toxin pretreatment of the animals indicating that a pertussis toxin-sensitive N-protein mediates both effects. Similar results were obtained with adenosine (not shown).

Fig. 1. Effect of (-)-N^6-phenylisopropyladenosine (PIA) on transmembrane action potential (upper traces) and force of contraction (lower traces) in guinea-pig auricles without (A) and with pretreatment with pertussis toxin (B; 150 µg/kg i.v.). Time of incubation with PIA was 5 min. Note that pretreatment with pertussis toxin prevented both the shortening of the action potential duration (APD_{90}) and the negative inotropic effect of PIA.

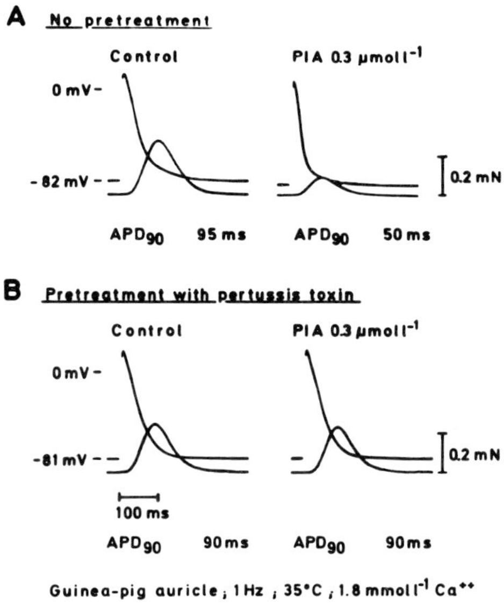

Since adenosine and m-cholinoceptor agonists like carbachol have quite similar effects on atrial force of contraction and on action potential and since carbachol has recently been shown to stimulate myocardial phosphoinositide turnover we investigated the effects of adenosine on the phosphoinositide turnover (table 1). The effects of carbachol were studied for comparison. Carbachol (1 µmol/l) increased IP_3 and decreased its precursor PIP_2. The effects were antagonized by atropine (10 µmol/l). In contrast, adenosine (100 µmol/l) or adenosine plus the adenosine receptor antagonist theophylline (500 µmol/l) did not affect the phosphoinositide turnover.

Table 1. Effects of adenosine (Ad, 100 µmol/l) and Ad plus theophylline (Theo, 500 µmol/l) on phosphoinositide turnover in rat electrically driven left auricles. Effects of carbachol (Carb, 1 µmol/l) and Carb plus atropine (Atr, 10 µmol/l) were studied for comparison. The products measured were inositol 1,4,5-trisphosphate (IP_3) and phosphotidylinositol 4,5-bisphosphate (PIP_2); n = 8-10 each. The numbers refer to means ± S.E.M (dpm/mg wet weight). Asterisks denote significant differences from control. Note that carbachol increased IP_3 and decreased PIP_2 whereas adenosine had no effect.

	Control	Ad	Ad plus Theo
IP_3	85 ± 12	86 ± 6	70 ± 7
PIP_2	663 ± 115	573 ± 161	557 ± 83

	Control	Carb	Carb plus Atr
IP_3	110 ± 9	174 ± 22*	94 ± 6
PIP_2	378 ± 54	196 ± 40*	337 ± 48

Effects of PIA on ventricular preparations

In particulate membrane preparations from right ventricles of guinea-pigs the effects of PIA on adenylate cyclase activity in the presence of isoprenaline (10 µmol/l) were studied with and without adenosine deaminase (ADA). The experiments were performed in the presence of 100 µmol/l GTP and 50 mmol/l NaCl. ADA increased basal (C_1 in fig. 2) as well as isoprenaline-stimulated (C_2 in fig. 2) adenylate cyclase activity by about 20 % indicating that a considerable amount of endogenous adenosine may be present in the assay and may attenuate adenylate cyclase activity.

Without ADA PIA did not attenuate isoprenaline-stimulated activity. In contrast, in the presence of ADA PIA concentration-dependently reduced isoprenaline-stimulated adenylate cyclase activity maximally by about 20 %, i.e. by the amount by which ADA increased the activity.

Fig. 2. Effects of (-)-N^6-phenylisopropyladenosine (PIA) in the presence of isoprenaline (Iso; 10 μmol/l) with and without adenosine deaminase (ADA; 1 μg/ml) on adenylate cyclase activity in a particulate membrane preparation from guinea-pig right ventricles. Asterisks denote significant differences from the corresponding isoprenaline values. Note that PIA decreased adenylate cyclase activity in the presence of ADA only.

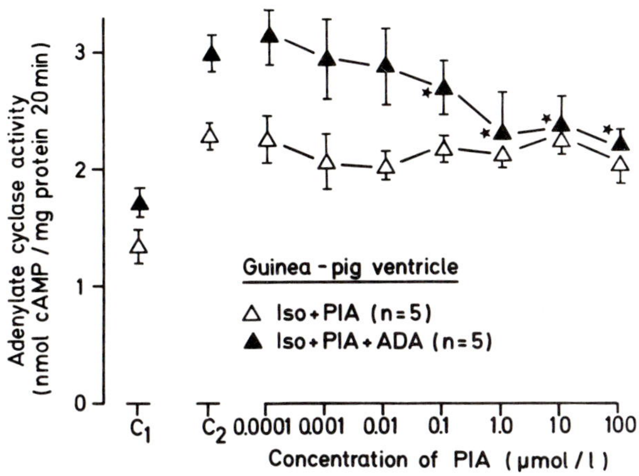

To relate the PIA-induced inhibition of adenylate cyclase activity to the negative inotropic effect we measured cAMP content and force of contraction in the same intact contracting papillary muscles. All experiments were performed in the presence of ADA. As illustrated in figure 3 ADA alone did not affect cAMP content or force of contraction. The positive inotropic effect of isoprenaline (0.01 μmol/l) was accompanied by an increase in cAMP content. Additionally applied PIA reduced force of contraction by about 40 % and cAMP content by about 20 % at 0.1 - 1 μmol/l. At higher concentrations of PIA (10-100 μml/l) force of contraction did not change further whereas cAMP increased to the isoprenaline value (10 μmol/l PIA) or even to higher values (100 μmol/l PIA), i.e. at these high concentrations PIA increased cAMP content of the papillary muscles.

In order to investigate whether the negative inotropic effect of PIA can be explained by a change in slow calcium inward current (I_{si}) we investigated the effects

Fig. 3. cAMP content (A) and force of contraction (B) in guinea-pig electrically driven papillary muscles in drug-free bathing solution (a, control), 30 min after addition of ADA (b, 1 µg/ml), 10 min after the addition of isoprenaline (c, Iso, 0.01 µmol/l) and 5 min after the additional application of various concentrations of PIA (d-k). Note that the "indirect" negative inotropic effect of PIA was accompanied by a decrease in cAMP content at low (g-h) concentrations and by an increase in cAMP at higher (i-k) concentrations of PIA.

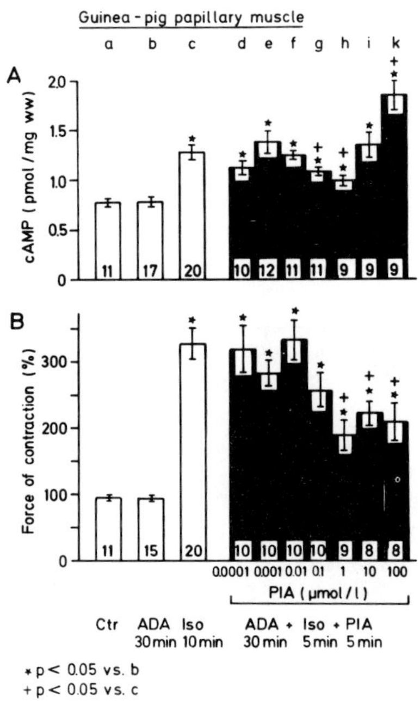

of PIA on dV/dt_{max} of slow action potentials (a measure of the slow calcium inward current). The negative inotropic effect of PIA was accompanied by a decrease of the isoprenaline-stimulated I_{si} (fig. 4). The effect of PIA was abolished by pretreatment with pertussis toxin indicating that a pertussis toxin sensitive N-protein is involved in the adenosine receptor-mediated effects in ventricular myocardium.

Fig. 4. Effect of isoprenaline (Iso; 0.01 µmol/l) and Iso plus (-)-N^6-phenylisopropyladenosine (PIA; 1 µmol/l) on calcium-dependent slow action potentials in untreated (A) and pertussis toxin-pretreated (B) guinea-pig electrically driven papillary muscles. Parameters: slow calcium-dependent action potentials (mV, upper traces), force of contraction (mN, middle traces) and maximal rate of depolarization (Vs^{-1}, lower traces). Note that both the negative inotropic effect of PIA and the attenuation by PIA of the Iso-induced increase in maximal rate of depolarization were abolished by pertussis toxin.

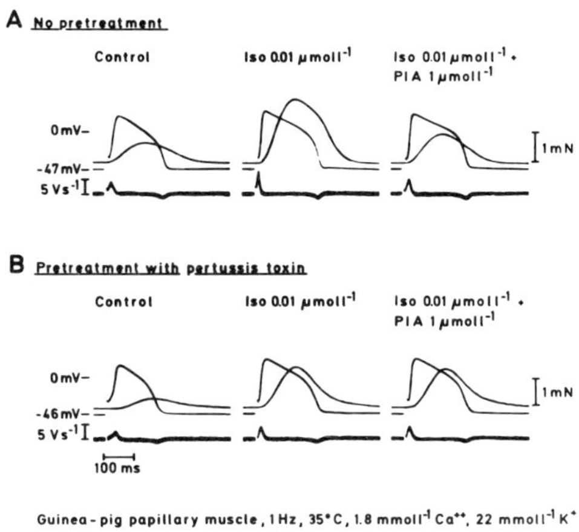

DISCUSSION

In atrial preparations the "direct" negative inotropic effects of adenosine or PIA were accompanied by a shortening of the action potential duration. Both, the negative inotropic and the action potential shortening effects were abolished by pretreatment with pertussis toxin. Since the "direct" negative inotropic effect of adenosine on atrial preparations occurred without changes in cAMP content (Endoh et al. 1983a, 1983b; Böhm et al. 1984; Brückner et al. 1985) and since the shortening of the action potential is due to an increase in K^+-conductance (Jochem & Nawrath 1983; Belardinelli & Isenberg 1983a; Böhm et al. 1984; Brückner et al. 1985) it is concluded that the negative inotropic effect in atrial preparations is due to the shortening of the action potential reducing the time available for Ca^{++} entry and thereby reducing force of contraction. The atrial adenosine receptor may directly be coupled to a K^+ channel via a pertussis toxin-sensitive N-protein as has also been suggested for the atrial m-cholinoceptor (Endoh et al. 1985; Pfaffinger et al. 1985; Sorota et al. 1985; Böhm et al. 1986; Schmitz et al. 1986). Activation of the phosphoinositide turnover does not appear to play a role in mediating the negative inotropic effect of adenosine. Although the m-cholinoceptor agonist carbachol increased phosphoinositide turnover it is unclear whether or not this effect contributes to the negative inotropic effect of m-cholinoceptor agonists (Brown & Brown Masters 1984).

In ventricular particulate membrane preparations PIA reduced the isoprenaline-stimulated adenylate cyclase activity by about 20 % in the presence of ADA only (and in the presence of 50 mmol/l NaCl and 100 μmol/l GTP). These results indicate that the small inhibitory effects of PIA on adenylate cyclase activity may be masked by endogenous adenosine. The same holds true for the small cAMP-decreasing effect at low concentrations of PIA in intact preparations observed in the present study because without ADA no decrease in cAMP content has previously been detected (Schmitz et al. 1985; Scholz et al. 1987). The small decrease in cAMP content is in accord with others (Schrader et al. 1977; Dobson 1983; Linden et al. 1985). However, in addition, at high concentrations of PIA we found an increase in cAMP content whereas force of contraction did not increase. This may be explained by the presence of adenylate cyclase stimulatory low affinity A_2 adenosine receptors in endothelial cells present in the guinea-pig ventricle (Schütz et al. 1986). Thus, the biphasic effect of PIA on cAMP content, but not on force of contraction, may be mediated by myocardial adenosine receptors which are coupled inhibitory to adenylate cyclase and which mediate at least in part the decrease in force of contraction and by endothelial A_2 adenosine receptors which are coupled in a stimulatory manner to adenylate cyclase and which are unrelated to myocardial force development.

Although the small decrease of the isoprenaline-induced elevation of cAMP content may at least in part explain the "indirect" negative inotropic effect of PIA , other mechanisms may also play a role. For instance, a direct decrease in protein kinase or increase in phosphatase activity can also explain the observed decrease in I_{si} which finally is responsible for the negative inotropic effect. If this is correct the ventricular adenosine receptor may be coupled via a pertussis toxin-sensitive N-protein not only to adenylate cyclase but also to other biochemical steps leading to less phosphorylation of the Ca^{++} channels and hence to a decrease in force of contraction.

ACKNOWLEDGEMENTS
This work was supported by the Deutsche Forschungsgemeinschaft.

REFERENCES

Anand-Srivastava, M.B., 1985, Archives of Biochemistry and Biophysics, 243, 439-446.
Baumann, G., Schrader, J. & Gerlach E., 1981, Circulation Research, 48, 258-266.
Belardinelli, L. & Isenberg, G., 1983a, American Journal of Physiology, 244, H734-H737.
Belardinelli, L. & Isenberg, G., 1983b, Circulation Research, 53, 287-297.
Berridge, M.J., Downes, C.P. & Hanley, M.R., 1982, Biochemical Journal, 206, 587-595.
Böhm, M., Brückner, R., Hackbarth, I., Haubitz, B, Linhart, R., Meyer, W., Schmidt, B., Schmitz, W. & Scholz, H., 1984, Journal of Pharmacology and Experimental Therapeutics, 230, 483-492.
Böhm, M., Brückner, R., Meyer, W., Nose, M., Schmitz, W., Scholz, H. & Starbatty, J., 1985, Naunyn-Schmiedeberg's Archives of Pharmacology, 331, 131-139.
Böhm, M., Brückner, R., Neumann, J., Schmitz, W., Scholz, H. & Starbatty J., 1986, Naunyn-Schmiedeberg's Archives of Pharmacology, 332, 403-405.
Brown, J.H. & Brown Masters, S., 1984, Trends in Pharmacological Sciences, 5, 417-419.
Brückner, R., Schmitz, W., Scholz, H., Schumacher, A., Erdmann, E., Krawietz, W. & Werdan, K., 1980, Naunyn-Schmiedeberg's Archives of Pharmacology, 315, 147-153.
Brückner, R., Fenner, A., Meyer, W., Nobis, T.-M., Schmitz, W. & Scholz, H., 1985, Journal of Pharmacology and Experimental Therapeutics, 234, 766-774.
Creba, J.A., Downes, C.P., Hawkins, P.T., Brewster, G., Michell, R.H. & Kirk, C.J., 1983, Biochemical Journal, 212, 733-747.
Dobson jr., J.G., 1978, Circulation Research, 43, 785-792.
Dobson jr., J.G., 1983, American Journal of Physiology, 245, H468-H474.
Drury, A.N. & Szent-Györgyi, A., 1929, Journal of Physiology, 68, 213-237.
Endoh, M., Maruyama, M. & Taira, N., 1983a, Journal of

Cardiovascular Pharmacology, 5, 131-142.

Endoh, M., Maruyama, M. & Taira, N, 1983b, In *Physiology and Pharmacology of Adenosine*, edited by J.W. Daly, Y. Kuroda, J.W. Phillis, H. Shimizu & M. Ui (Raven press, New York), p. 127-140.

Endoh, M., Maruyama, M. & Ijima, T., 1985, *American Journal of Physiology*, 249, H309-H320.

Huang, M. & Drummond, G.I., 1976, *Biochemical Pharmacology*, 25, 2713-2719.

Jochem, G. & Nawrath, H., 1983, *Experientia*, 39, 1347-1349.

Linden, J., Hollen, C.E. & Patel, A., 1985, *Circulation Research*, 56, 728-735.

Nawrath, H., Jochem, G. & Sack, U., 1985, In *Adenosine: Receptors and Modulation of Cell Function*, edited by V. Stefanovich, K. Rudolphi, P. Schubert (IRL Press, Oxford), p. 323-340.

Pfaffinger, P.J., Martin, J.M., Hunter, D.D., Nathanson, N.M. & Hille, B., 1985, *Nature*, 317, 536-538.

Schmitz, W., Böhm, M., Brückner, R., Burmann, H., Meyer, W., Neumann, J., Nose, M. & Scholz H., 1985, In *Adenosine: Receptors and Modulation of Cell Function*, edited by V. Stefanovich, K. Rudolphi, P. Schubert (IRL Press, Oxford), p. 295-303.

Schmitz, W., Brückner, R. & Neumann, J., 1986, *Naunyn-Schmiedeberg's Archives of Pharmacology*, 332, R 46.

Schmitz, W., Scholz, H., Scholz, J., Steinfath, M., Lohse, M., Puurunen, J. & Schwabe, U., 1987a, *European Journal of Pharmacology*, 134, 377-378.

Schmitz, W., Scholz, H., Scholz, J. & Steinfath, M., 1987b, *European Journal of Pharmacology*, 140, 109-111.

Scholz, H., Böhm, M., Brückner, R., Neumann, J. & Schmitz, W., 1987, In *Topics and Perspectives in Adenosine Research*, edited by E. Gerlach & B.F. Becker (Springer-Verlag, Berlin, Heidelberg), p. 369-382.

Schrader, J., Baumann, G. & Gerlach, E., 1977, *Pflügers Archives*, 372, 29-35.

Schütz, W., Freissmuth, M., Hausleithner, V. & Tuisl, E., 1986, *Naunyn-Schmiedeberg's Archives of Pharmacology*, 333, 156-162.

Sorota, S., Tsuji, Y., Tajima, T. & Pappano, A.J., 1985, *Circulation Research*, 57, 748-758.

RELEASE OF PROSTACYCLIN FROM ENDOTHELIAL CELLS BY ADP AND ATP : MECHANISM OF ACTION

Boeynaems, J.M., Raspe, E., Pirotton, S., Demolle, D., Van Coevorden, A. & Erneux, C.

Institute of Interdisciplinary Research, School of Medicine, Université Libre de Bruxelles, Campus Erasme, 1070 - Brussels, Belgium.

Extracellular ADP and ATP have been shown in the last years to stimulate the release of 2 mediators from arterial endothelial cells : "endothelium-derived relaxing factor" (EDRF) (Burnstock and Kennedy, 1986), a vasodilator recently identified as NO (Palmer et al., 1987) and prostacyclin (PGI_2), a potent inhibitor of platelet aggregation (Boeynaems and Galand, 1983; Pearson et al., 1983; Van Coevorden and Boeynaems, 1984; Hellewell and Pearson, 1984). Studies of agonist specificity have indicated that these 2 actions are mediated by P_2-receptors of the P_{2y} subtype (Martin et al., 1985; Needham et al., 1987). ATP and ADP, as well as their phosphorothioate analogs, ATPγS and ADPβS, are equipotent and equiactive in releasing PGI_2, while the non-hydrolysable analog APPNP is active though less potent than ATP and the P_{2x}-selective agonist APCPP is completely inactive, like AMP and adenosine. Studies with the Ca^{++} ionophore A23187 have indicated that the cytoplasmic concentration of Ca^{++} plays a crucial role in the control of arachidonic acid mobilization and prostaglandin synthesis by endothelial as well as by other cells (Weksler et al., 1978; Adams Brotherton and Hoak, 1982). Our recent studies have demonstrated that ADP and ATP induce a rapid and transient accumulation of Ins 1,4,5-trisphosphate (IP_3) in bovine aortic endothelial cells, suggesting that $P_{2}y$ receptors are coupled to a phospholipase C hydrolysing phosphatidyl inositol bisphosphate (PIP_2) (figs 1 and 2). Similar findings have been made in rat hepatocytes (Charest et al., 1985) and H-35 hepatoma cells (Horstman et al., 1986). We have obtained preliminary evidence that the coupling between P_2-receptors and the PIP_2-phospholipase C involves a GTP binding protein : the accumulation of IP_3 in response

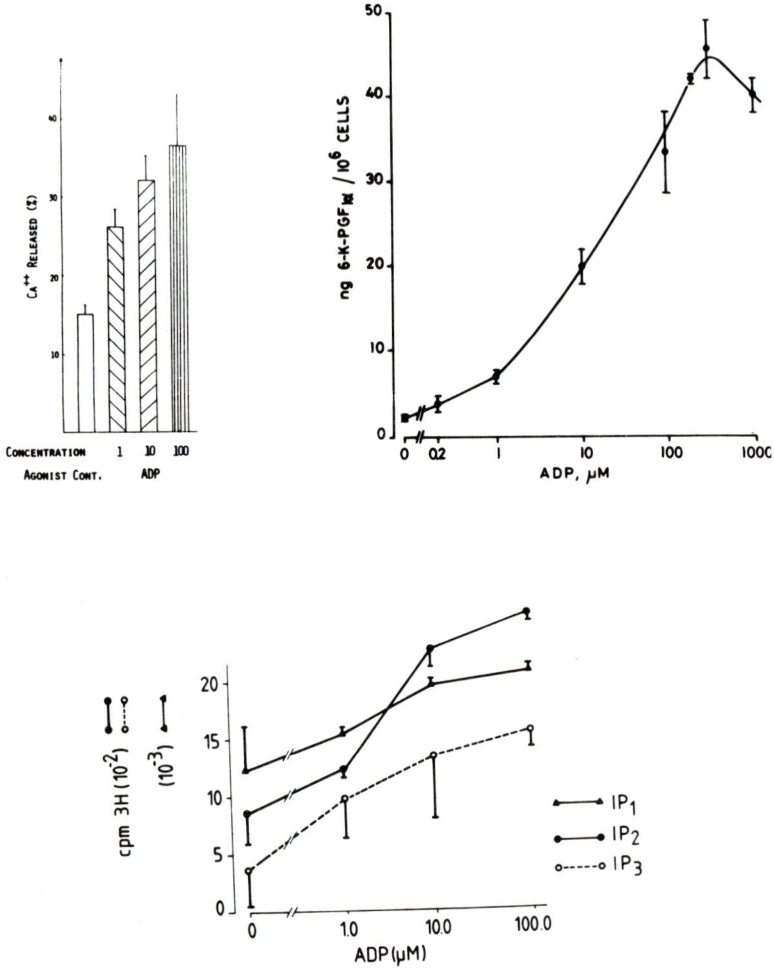

Figure 1.

Concentration-action curves characterizing the effect of adenine nucleotides on the accumulation of Ins 1,4,5-P_3 (IP_3), the outflow of $^{45}Ca^{++}$ and the release of PGI_2, in bovine aortic endothelial cells. PGI_2 was measured as its stable degradation product, prostaglandin 6-keto-$F_{1\alpha}$ (6-K-$PGF_{1\alpha}$).

to ATP was indeed largely decreased in aortic endothelial cells which had been treated with pertussis toxin, but not cholera toxin (fig. 3) (Pirotton et al., submitted ; Boeynaems et al., 1987). As expected, ADP and ATP induced an increase of $^{45}Ca^{++}$ efflux from prelabelled bovine aortic endothelial cells (fig. 1), which is partially maintained in a Ca^{++}-free medium, suggesting the involvement of an intracellular Ca^{++} store, like the endoplasmic reticulum which is the target of Ins 1,4,5-P_3 (Berridge and Irvine, 1984; Irvine et al., 1986). Measurements of the fluorescence emission by bovine aortic endothelial cells loaded with quin-2 demonstrated that ATP and ADP induce a rapid and transient increase of the cytoplasmic concentration of free Ca^{++} from 90 nM to values between 200 and 600 nM (fig. 2; Luckhoff and Busse, 1986; Hallam and Pearson, 1986). Similar observations have been made in rat hepatocytes with the quin-2 method (Charest et al, 1985; Sistare et al., 1985); furthermore, by means of aequorin measurements in single hepatocytes, Cobbold et al. (1987) have demonstrated that the Ca^{++} response to ATP and ADP consists of repetitive transient rises lasting ± 10 sec.

Our studies of agonist specificity, time course and concentration-action curves are consistent with the existence of a causality link between the various events triggered by adenine nucleotides in aortic endothelial cells: IP_3 accumulation, cytosolic Ca^{++} increase and PGI_2 release. Indeed concentration-action curves for the 3 events span the 1-100 μM range of concentration (fig. 1). By comparison, the hepatocytes are much more sensitive to adenine nucleotides, since half-maximal stimulations of IP_3 accumulation and phosphorylase activity are obtained at submicromolar concentrations (Charest et al., 1985). However it is known that the plasma concentration of (ADP + ATP) reaches ± 20 μM following platelet activation by thrombin. It is therefore likely that the responses to adenine nucleotides described here could occur in vivo when platelets aggregate on the site of an endothelial injury : this release of PGI_2 might have the physiological function of limiting the extent of platelet aggregation and helping to localize thrombus formation to areas of endothelium damage. As shown on fig. 2, the increases of IP_3 and cytosolic Ca^{++} induced by ATP or ADP are rapid in onset (maximum between 15 and 30 sec) and transient (return to the baseline in 1-2 min); the increase in PGI_2 release is detectable within 2 min of ADP addition and does not last much more than 5 min. Finally, PGI_2 release and IP_3

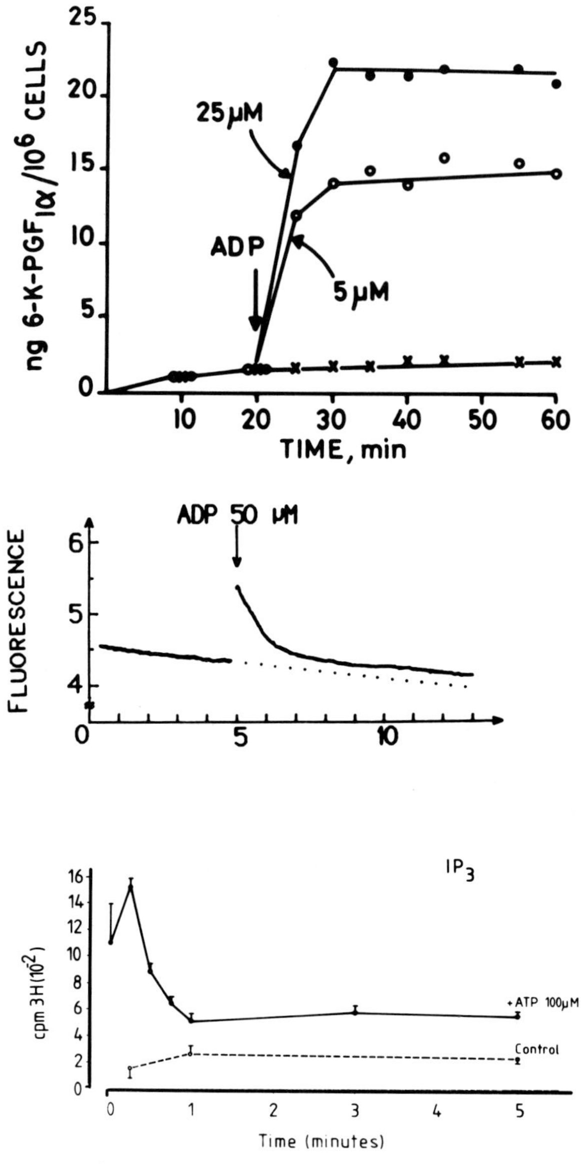

Figure 2.

Time course of IP_3 accumulation, increase of cytosolic Ca^{++} and release of PGI_2, in bovine aortic endothelial cells stimulated by adenine nucleotides.

accumulation are characterized by the same agonist specificity : ATP = ATPγS = ADP = ADPβS > APPNP > APCPP, AMP = Adenosine = 0. In conclusion all the available data suggest a very simple mechanism : the burst of PGI_2 synthesis in response to adenine nucleotides would result from the rapid and transient increase of cytosolic Ca^{++}, due to Ca^{++} release from the endoplasmic reticulum by Ins 1,4,5-P_3, possibly via the activation of a Ca^{++}-dependent phospholipase A_2.

However, recent and preliminary data do not fit well with such a straightforward mechanism. Treatment of bovine aortic endothelial cells with pertussis toxin inhibited the accumulation of IP_3 but potentiated the release of PGI_2 induced by ATP (fig. 3). Endothelial cells loaded with quin-2, in conditions known to produce a significant buffering of cytoplasmic Ca^{++}, produced more PGI_2 in response to ATP than untreated cells (table 1). This paradoxical observation indicates that quin-2 is not always an innocent indicator and may exert pharmacological actions. A similar treatment with quin-2 inhibited the aggregation of platelets induced by arachidonic acid (Rao et al., 1986) and the release of von Willebrand factor from human umbilical vein endothelial cells, stimulated by histamine (Hamilton and Sins, 1987). These data raise the alternative hypothesis that P_2-receptors in endothelial cells might control PGI_2 production by mechanisms other than Ins 1,4,5-P_3 and cytosolic Ca^{++} increase. It was recently shown that α_1-adrenergic receptors in FRTL5 thyroid cells are coupled to both a phospholipase A_2 releasing free arachidonic acid and to a phospholipase C generating Ins 1,4,5-P_3 (Burch et al., 1986). Nakashima et al. (1987) have demonstrated recently that a phospholipase A_2 in platelets is directly controlled by a GTP-binding protein. This hypothesis is currently being tested in aortic endothelial cells.

This work was performed under contract of the "Ministère de la Politique Scientifique" ("Action Concertée") and supported by grants from Duphar and the "Fondation Médicale Tournay-Solvay". C. Erneux and E. Raspe were supported by IRSIA.

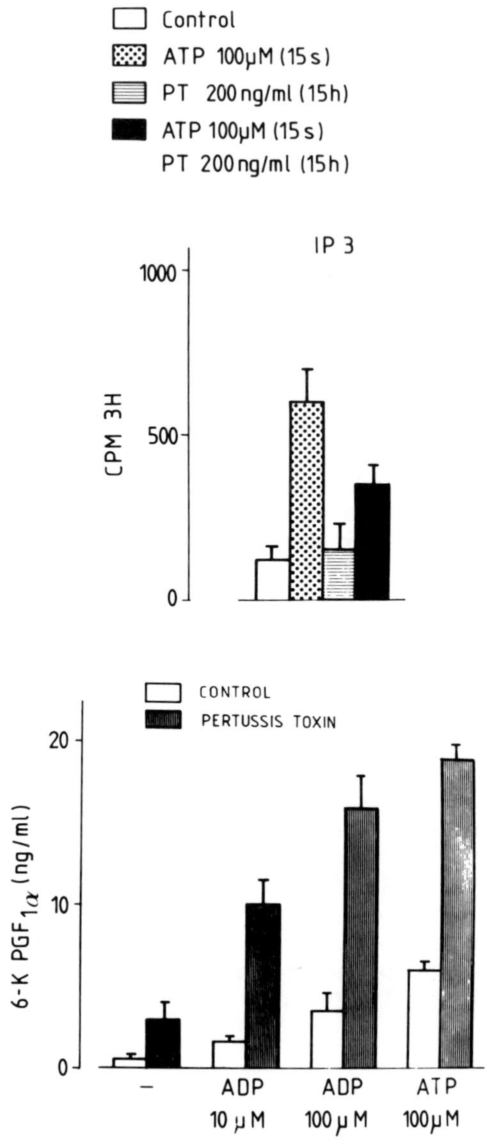

Figure 3.

Dual effect of pertussis toxin on the responsiveness of bovine aortic endothelial cells to adenine nucleotides: decrease of IP_3 accumulation and increase of PGI_2 release.

Table 1. Potentiation by quin-2 loading of the stimulation by ATP of PGI_2 release from bovine aortic endothelial cells.

µM quin-2AM	-	ATP (10 µM)
0	1.9 ± 0.9	4.8 ± 1.2
5	1.6 ± 0.6	10.0 ± 3.0
24	1.7 ± 0.6	16.0 ± 4.0
120	2.1 ± 1.4	15.0 ± 3.0

Bovine aortic endothelial cells were incubated for 60 min in the presence of quin-2 AM at various concentrations. Then, they were incubated for 20 min with or without ATP. Results represent the release of PGI_2 during that incubation (ng 6-K-$PGF_{1\alpha}$ /ml; mean ± SD of 6 determinations).

REFERENCES

Adams Brotherton, A.F. & Hoak, J.C., 1982, Proceedings of the National Academy of Sciences USA, 79, 495.

Berridge, M.J. & Irvine, R.F. 1984, Nature, 312, 315.

Boeynaems, J.M. & Galand, N., 1983, Biochemical and Biophy sical Research Communications, 112, 290.

Boeynaems, J.M., Pirotton, S., Van Coevorden, A., Raspe, E., Demolle, D. & Erneux, C., 1987, Journal of Receptor Research, in press.

Burch, R.M., Luini, A & Axelrod, J., 1986, Proceedings of the National Academy of Sciences, USA, 83, 7201.

Burnstock, G. & Kennedy, C., 1986, Circulation Research, 58, 319.

Charest, R., Blackmore, P.F. & Exton, J.H., 1985, Journal of Biological Chemistry, 260, 15789.

Cobbold, P., Cuthbertson, R., Woods N. & Publicover, S., 1987, Journal of Receptor Research, in press.

Hallam, T.J. & Pearson, J.D., 1986, FEBS Letters, 207, 95.

Hamilton, K.K. & Sims, P.J. 1987, Journal of Clinical Investigation, 79, 600.

Hellewell, P.G. & Pearson, J.D., 1984, British Journal of Pharmacology, 83, 457.

Horstman, D.A., Tennes, K.A. & Putney, J.W., 1986, FEBS Letters , 202, 189.

Irvine, R.F., Letcher, A.J., Lander, D.J. & Berridge, M.J., 1986, Biochemical Journal, 240, 301.

Luckoff, A. & Busse, R., 1986, Journal of Cell Physiology, 126, 414.

Martin, W., Cusack, N.J., Carleton, J.S. & Gordon, J.L., 1985, European Journal of Pharmacology, 108, 295.

Nakashima, S., Tohmatsu, T., Hattori, H., Suganuma, A. & Nozawa, Y., 1987, Journal of Biochemistry, 101, 1055.

Needham, L., Cusack, N.J., Pearson, J.D. & Gordon, J.L., 1987, European Journal of Pharmacology, 134, 199.

Palmer, R.M.J., Ferridge, A.G. & Moncada, S., 1987, Nature 327, 524.

Pearson, J.D., Slakey, L.C. & Gordon, J.L., 1983, Biochemical Journal, 214, 273.

Rao, G.H.R., Peller, J.D., Semba, C.P. & White, J.G., 1986, Blood, 67, 354.

Sistare, F.D., Picking, R.A. & Haynes, R.C., 1985, Journal of Biological Chemistry, 260, 12744.

Van Coevorden, A. & Boeynaems, J.M., 1984, Prostaglandins 27, 615.

Weksler, B.B., Ley, C.W. & Jaffe, E.A., 1978, Journal of Clinical Investigation, 62, 923.

ADENOSINE MEDIATED REGULATION OF HEPATIC VASCULATURE

W.W. LAUTT

HEPATORENAL RESEARCH UNIT
DEPARTMENT OF PHARMACOLOGY AND THERAPEUTICS
FACULTY OF MEDICINE
UNIVERSITY OF MANITOBA
WINNIPEG, MANITOBA
CANADA, R3E 0W3

OVERVIEW

Although the liver constitutes only 2.5% of body weight, it receives 25% of the cardiac output. Of the total hepatic blood flow (30 ml/min/Kg body weight), 20-33% is supplied by the hepatic artery. About two-thirds of the hepatic blood supply is portal venous blood that has drained from the extrahepatic splanchnic organs (stomach, intestine, pancreas, spleen, omentum). The liver does not directly control portal blood flow. Changes in intrahepatic vascular resistance will affect portal venous pressure but the most dramatic acute vasoconstriction in liver induced, for example by hepatic sympathetic nerves, will double portal venous pressure without changing portal flow. The liver therefore must accept whatever volume of portal flow is sent to it by the splanchnic organs. The hepatic artery is the sole means by which the liver can directly control its total hepatic blood flow. A detailed review (Greenway & Lautt 1987) and a more abbreviated conceptual review (Lautt & Greenway 1987) should be consulted for details of hepatic vascular regulation including the areas of exchange of substances between blood and parenchymal cells and the extremely important capacitance (blood reservoir) functions of the liver.

Hepatic Artery: Prior to 1977, it was generally accepted that the blood flow in the hepatic artery was regulated according to the metabolic demands of the liver, a regulatory system that was analogous to that of all other organs. At that time (Lautt 1977) it was shown that reduction of the oxygen supply to the liver produced by isovolemic hemodilution did not lead to the expected dilation of the hepatic artery. Changes in hepatic arterial flow were related not to blood gases in the hepatic artery, portal vein, or hepatic venous outflow but were rather related to portal flow per se in an inverse manner. Subsequent studies confirmed that elevation or depression of liver metabolism by the use of dinitrophenol or SKF525A also did not result in the anticipated changes in hepatic arterial flow (Lautt 1980). The relationship between liver blood flow and liver metabolism was reviewed (Lautt 1983) and the conclusion was that liver parenchymal cell metabolism does not directly regulate hepatic arterial vascular tone.

Intrinsic regulatory mechanisms consist of two forms of regulation both of which are mediated by adenosine and both are designed to maintain total blood flow per se (not oxygen delivery) at a constant level. The importance of maintaining total blood flow constant has been previously discussed (Lautt & Greenway 1987) and includes important factors such as maintenance of intrahepatic and portal venous pressures as well as having extremely major consequences on hormone clearance rates. Clearance by the liver of many compounds is blood flow limited therefore it is important for overall homeostasis that blood flow changes in the liver be minimized despite possibly large transient changes in portal venous flow. This flow regulation is achieved by the two intrinsic regulatory systems, the hepatic arterial buffer response (HABR) and autoregulation of the hepatic artery.

Hepatic Arterial Buffer Response (HABR): The hepatic arterial buffer response (Lautt 1981) defines the relationship between portal venous flow and hepatic arterial flow whereby a change in portal flow results in the inverse change in hepatic arterial flow. The mechanism and role of this response has recently been reviewed (Lautt 1985). Various hypotheses were initially tested in order to determine the mechanism of the hepatic arterial buffer response (HABR) and eventually the only hypothesis that could not be eliminated was based on washout of some vasodilator compound from the region of the hepatic arterial resistance vessels (Lautt 1981).

The current hypothesis is explained in Figure 1.

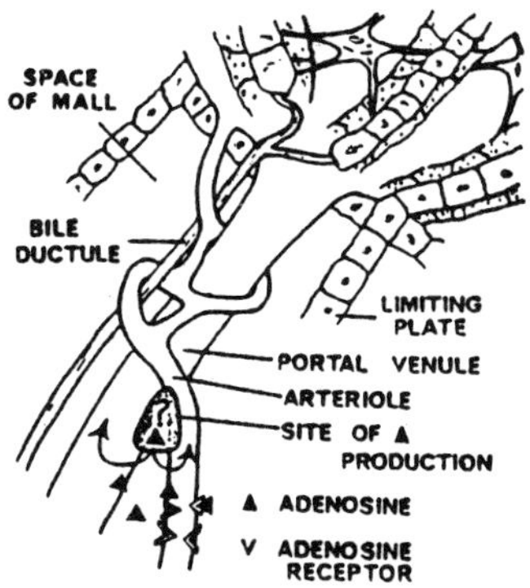

Figure 1. HABR Hypothesis. Adenosine washout hypothesis: The area shown is the portal triad which represents the vascular stalk leading into the center of each acinus (100,000 acini per liver). The terminal branches of the hepatic arteriole, portal venule and bile ductule lie within an enclosed space delimited by a limiting plate of cells. The fluid surrounding these vessels and contained within the limiting plate comprises the Space of Mall. The data are consistent with the hypothesis that adenosine is continuously secreted into the Space of Mall (independent of general parenchymal cell oxygen supply or demand) and that the local concentration of adenosine determines the tone of the hepatic artery. The adenosine can be washed away into the portal vein. The hepatic arterial buffer response is accounted for by, for example, reduced portal blood flow washing away less adenosine and the accumulated adenosine leading to dilation of the hepatic artery. Figure 1 reproduced from Lautt 1985.

Several criteria have been fulfilled to support the hypothesis that the HABR is mediated by adenosine. First, adenosine is a dilator of the hepatic artery that can result in elevation of arterial conductance to almost three times basal levels (Lautt & Legare 1985). Second, the hypothesis proposes that portal blood flow must be able to wash adenosine away from a common pool which has access to the arterial resistance vessels. Therefore, portal flow must have access to the fluid surrounding the hepatic artery. This has been shown by the observation that vasoactive compounds infused into the portal vein are capable of diffusing to the hepatic artery and producing vascular responses that are one half to one third the magnitude of the same dose infused directly into the hepatic artery (Greenway & Lautt 1987; Lautt et al. 1984; Richardson & Withrington 1981).

In order to establish other criteria for this hypothesis, it was necessary to develop a vascular preparation in which reproducible and easily quantitated buffer responses could be measured. The methodology used has been described and evaluated (Lautt et al. 1985) and involves elimination of all portal supply except that provided by the superior mesenteric artery (SMA).

Hepatic arterial and SMA flows are measured using electromagnetic flowprobes. Complete occlusion of the SMA rapidly reduces portal flow to zero and the hepatic artery undergoes immediate dilation independent of changes in blood pressure. This dilation is the hepatic arterial buffer response. There are various means of quantitating the buffer response, however the most consistent means of expression is the buffer capacity calculated as the change in hepatic arterial flow divided by the change in portal flow expressed as a percent, that is, the proportion of the change in portal flow compensated by the hepatic arterial flow. The buffer capacity in this type of preparation is only 25% and while we do not believe that this is a normal physiological value, it nevertheless allows mechanistic studies to be carried out.

By the use of this methodology we were able to fulfill two additional criteria. Dipyridamole (given IV as a bolus, 400 ug/Kg, followed by a constant infusion, 136 ug/Kg/min) resulted in mild but significant potentiation of the vasodilator effect of a standard intra-arterial (i.a.) test dose of adenosine (0.04 mg/Kg/min). In addition, the hepatic arterial buffer response rose from control levels of 23% to 34% in the presence of dipyridamole. All other means of quantitating the buffer response also indicated that dipyridamole produced a mild potentiation of both the vasodilator effects of exogenously administered adenosine and of the buffer response. The criterion of antagonism of the buffer response with adenosine antagonists has been achieved by the use of 3 antagonists, 1-methyl-3-isobutylxanthine (Lautt et al. 1985), 8-(p-sulfophenyl)theophylline (Ezzat & Lautt 1987) and 8-phenyltheophylline (Lautt & Legare 1985). These antagonists have been shown to antagonize both the vasodilator response to exogenous adenosine and the buffer response. Of these three antagonists, 8-phenyltheophylline (8-PT) was clearly superior for our needs. Nevertheless, it has proven a difficult compound to use with whole animal in vivo experiments because of the low solubility, variable dosage required and variable duration of action. In each individual experiment a control buffer response and a test hepatic arterial response to intraportal adenosine and to intraportal isoproterenol is obtained. 1 mg/Kg of 8-PT is then administered and the tests are repeated. If blockade of adenosine has not occurred the 8-PT dose is doubled and the blockade again tested. This procedure is continued until a full blocking dose is obtained and subsequent tests with the buffer response are followed by repeated infusions of adenosine to confirm that the blockade remained. Effective blocking doses of 8-PT administered directly into the hepatic artery range from 1-16 mg/Kg. This highly variable dosage has also been found to occur for similar studies of autoregulation in the SMA (Lautt 1986). The duration of blockade with 8-PT has also proven problematic in that blockade may last between 15 and 30 minutes and during recovery from blockade it is common to see the response to infused adenosine and the buffer response potentiated. In these whole animal experiments we have also found it essential to verify in each animal that the blockade is selective and that the vasodilator response to isoproterenol is not impaired. Deterioration of the vascular preparation is thus precluded as a confounding factor in our interpretation. 8-PT is a competitive antagonist of exogenous adenosine-induced vasodilation of the hepatic artery and the dose related antagonism by 8-PT of the response to adenosine and the buffer response are roughly in parallel (Lautt & Legare 1985).

Autoregulation: The second intrinsic regulation of the hepatic artery is termed autoregulation. Autoregulation has classically been thought to be mediated by a myogenic response whereby an increase in arterial pressure leads to a stretch of the arterial resistance vessels and a subsequent vasoconstriction of these vessels. We tested the hypotheses that not only could adenosine concentrations be regulated in the Space of Mall by washout of adenosine into the portal blood but that similar washout could occur into the arterial blood stream thereby accounting for autoregulation (Ezzat & Lautt 1987). A micrometer controlled screwclamp was used to decrease hepatic arterial pressure. This resulted in a vasodilation of the hepatic artery indicating the presence of autoregulation. 8-PT administration resulted in dose- related suppression of several indices of autoregulation as well as the buffer response and the response to exogenous adenosine.

Adenosine as a Modulator of Constrictor Agents: It has previously been shown (Greenway et al. 1967) that the sympathetic nerves result in constriction of the hepatic artery which results in a reduced hepatic arterial flow. In addition, contraction of hepatic venous sphincters leads to an elevation of portal venous pressure. We tested the hypothesis that adenosine not only served as a vasodilator of the hepatic artery but could also antagonize the effects of vasoconstrictor stimuli including hepatic nerve stimulation. Intra-arterial infusions of adenosine, in concentrations that produced dilations similar to that seen during the buffer response, inhibited arterial vasoconstriction induced by intraportal administrations of norepinephrine, vasopressin, angiotensin, and hepatic nerve stimulation (Lautt & Legare 1986). The data are consistent with the hypothesis that adenosine causes nonselective inhibition of vasoconstrictor influences on the hepatic artery and that the neurally induced vasoconstriction is antagonized by a postsynaptic effect and does not involve interference with norepinephrine release. It is not, however, clear at this time whether endogenous levels of adenosine are able to be elevated sufficiently by physiological stimuli in order to serve as a modulator of neural and humoral vasoconstrictors.

Portal Vascular Resistance: In constrast to the hepatic artery, the intrahepatic portal resistance vessels are not affected by even large doses of adenosine. Basal tone was not altered nor was there seen to be any antagonism of vasoconstrictor effects of nerve stimulation, norepinephrine or angiotensin infusions (Lautt & Legare 1986). The data are consistent with the hypothesis that the smooth muscle that regulates portal pressure does not contain adenosine receptors. Recent studies (Lautt et al. 1987) have shown that these vasoconstrictor substances produce their effects on portal venous pressure by vasoconstriction of sphincters localized in the hepatic veins rather than presinusoidally in the portal venules as had previously been assumed. The notable difference in adenosine responsiveness between the hepatic artery and portal vein is, teleologically, a convenient phenomenon. The adenosine within the Space of Mall that affects the hepatic arteries does not appear to derive from normal hepatic parenchymal cells. Adenosine that might be released from hepatic parenchymal cells will enter the sinusoidal space and be washed away into the systemic circulation. Because of the anatomy of the hepatic vasculature with its unique acinar flow arrangement (Rappaport 1973) any substances released by parenchymal cells will be washed away from the arterial resistance vessels and will not have access to these vessels. This anatomical arrangement explains why metabolic activity of the parenchymal cells cannot affect the hepatic blood flow directly. Since the hepatic venous sphincters that control intrahepatic and portal pressure are located downstream from the parenchymal cells, metabolites entering the sinuoidal blood will bathe these sphincters. It is clear that the parenchymal cells of the liver release substantial quantities of adenosine and it is therefore essential that the sphincters not be sensitive to the variable release of adenosine from parenchymal tissue.

Conclusions: Data are consistent with the following hypothesis. Adenosine is secreted into the Space of Mall at a constant rate and independent of hepatic parenchymal cell metabolic status. The concentration of adenosine in the Space of Mall is regulated by washout into the portal blood and into the hepatic artery thus accounting for the mechanism of the hepatic arterial buffer response and arterial autoregulation respectively. Adenosine produces a nonselective antagonism of neural and humoral vasoconstrictors of the hepatic artery but is without effect on the hepatic venous sphincters which control portal venous pressure. Adenosine in the liver must be considered to exist in two independent pools. The adenosine released into the Space of Mall is essentially oxygen independent and is probably restricted to a very small fluid compartment in the vicinity of the arterial resistance vessels and well upstream from hepatic parenchymal cells. The adenosine regulation that occurs in hepatic parenchymal cells is probably oxygen related but when released by the liver is without direct vascular consequences on the hepatic circulation.

Ezzat, W.R. & Lautt, W.W. 1987 Hepatic arterial pressure-flow autoregulation is adenosine mediated. American Journal of Physiology, 252, H836-H845.

Greenway, C.V. & Lautt, W.W. 1987 The hepatic circulation. In Handbook of Physiology, American Physiological Society, (in press).

Greenway, C.V., Lawson, A.E. & Mellander S. 1967 The effects of stimulation of the hepatic nerves, infusions of noradrenaline and occlusion of the carotid arteries on liver blood flow in the anesthetized cat. Journal of Physiology, 192, 21-41.

Lautt, W.W. 1977 Control of hepatic and intestinal blood flow: effects of isovolaemic haemodilution on blood flow and oxygen uptake in the intact liver and intestines. Journal of Physiology, 265, 313-326.

Lautt, W.W. 1980 Control of hepatic arterial blood flow: independence from liver metabolic activity. American Journal of Physiology, 239, H559-H564.

Lautt, W.W. 1981 Role and control of the hepatic artery. In Hepatic Circulation in Health and Disease, (Raven

Press, NY).

Lautt, W.W. 1983 Relationship between hepatic blood flow and overall metabolism: the hepatic arterial buffer response. Federation Proceedings, 42, 1662-1666.

Lautt, W.W. 1985 Mechanism and role of intrinsic regulation of hepatic arterial blood flow: the hepatic arterial buffer response. American Journal of Physiology, 249, G549-G556.

Lautt, W.W. 1986 Autoregulation of superior mesenteric artery is blocked by adenosine antagonism. Canadian Journal of Physiology and Pharmacology, 64, 1291-1295.

Lautt, W.W. & Greenway, C.V. 1987 Conceptual review of the hepatic vascular bed. Hepatology, (in press).

Lautt, W.W., Greenway, C.V., & Legare, D.J. 1987 Effect of hepatic nerves, norepinephrine, angiotensin, elevated central venous pressure on postsinusoidal resistance sites and intrahepatic pressures in cats. Microvascular Research, 33, 50-61.

Lautt, W.W. & Legare, D.J. 1985 The use of 8-phenyl-theophylline as a competitive antagonist of adenosine and an inhibitor of the intrinsic regulatory mechanism of the hepatic artery. Canadian Journal of Physiology and Pharmacology, 63, 717-722.

Lautt, W.W. & Legare, D.J. 1986 Adenosine modulation of hepatic arterial but not portal venous constriction induced by sympathetic nerves, norepinephrine, angiotensin, and vasopressin in the cat. Canadian Journal of Physiology and Pharmacology, 64, 449-454.

Lautt, W.W., Legare, D.J., & d'Almeida, M.S. 1985 Adenosine as putative regulator of hepatic arterial flow (the buffer response). American Journal of Physiology, 248, H331-H338.

Lautt, W.W., Legare, D.J., & Daniels, T.R. 1984 The comparative effect of administration of substances via the hepatic artery or portal vein on hepatic arterial resistance, liver blood volume, and hepatic extraction in cats. Hepatology, 4, 927-932.

Rappaport, A.M. 1973 The microcirculatory hepatic unit. Microvascular Research, 6, 212-228.

Richardson, P.D. & Withrington, P.G. 1981 Liver blood flow 2. Effects of drugs and hormones on liver blood flow. Gastroenterology, 81, 356-375.

EFFECTS OF ADENOSINE ANALOGS ON RENAL HEMODYNAMICS AND RENIN RELEASE

H. Osswald

Goedecke Research Institute,
Mooswaldallee 1-9 7800 Freiburg,
West Germany

PHYSIOLOGICAL BACKGROUND OF THE RENAL ADENOSINE ACTIONS

Adenosine elicits vasoconstriction in the kidney. In order to understand this unusual vascular response it is necessary to analyze how blood flow and metabolic work are coupled in the kidney. In brain, heart and skeletal muscle blood flow increases when the metabolic rate of the tissue is enhanced. Since a rise in blood flow provides the tissue with an increase in supply with oxygen and energy substrates, a general concept of a metabolic control of organ blood flow was developed.

In the kidney the arterio-venous difference of oxygen is only 10 %. It follows that the kidney can extract more oxygen (and substrates) without increasing renal blood flow. Most of the blood flowing into the kidney passes through the glomeruli where the glomerular ultrafiltrate is formed. The glomerular filtration rate (GFR) represents a large leak in the circulatory system: 120 ml of ultrafiltrate per minute are leaving the vascular space in man. The tubular system reabsorbs more than 99 % of that filtrate before the final urine is formed. The major fraction of the renal oxygen consumption is used for active sodium transport powered by the Na, K-ATPase (see for ref. Klahr & Hammermann, 1985) while the tubular reabsorption of most of the other electrolytes and solutes is controlled by the active sodium transport. Thus, the glomerular filtration rate generates the work load of the renal tubules. As the glomerular filtration rate is highly dependent on renal blood flow we have in the kidney the reverse coupling between flow and metabolism compared to the heart: in the kidney flow controls metabolism, whereas in the heart metabolism controls flow. If we apply to the kidney the general principle of the metabolic control of organ blood flow, a vasodilating metabolite is unlikely to control renal blood flow because a positive feedback would be initiated: increase in GFR ⟶ increase in filtered NaCl ⟶ increase in tubular electrolyte transport ⟶ increase in oxygen consumption ⟶ vasodilation ⟶ increase in GFR. Such a positive feedback could not contribute to the known constancy of renal blood flow and glomerular filtration rate.

However, if we hypothesize that the increased metabolic rate in the kidney generates a vasoconstrictive metabolite, a homeostatic control of renal hemodynamics is quite possible. The idea that adenosine can in fact function in the kidney as a mediator of the metabolic control of renal function was discussed in more detail in previous papers (Osswald et al., 1980; Osswald, 1983). The juxtaglomerular apparatus where every nephron makes contact with its own glomerulus is the morphological substrate for hemodynamic responses to tubular events. The tubulo-glomerular feedback is the vascular response (afferent arteriolar vasoconstriction and reduced renin secretion) to an increased NaCl load to the tubules. Adenosine, formed by the tubular epithelium, is a mediator in the tubulo-glomerular feedback mechanism, according to Osswald et al. (1980).

The background of the physiological significance of the renal response to adenosine makes it very interesting to characterize by pharmacological means the mechanisms and receptors that are involved in the renal actions of adenosine.

There are four different structures in the kidney that can be influenced by adenosine, the renal vasculature, the renin containing cells, the tubular epithelium and the adrenergic nerve endings.

HEMODYNAMIC EFFECTS

The first suggestion of a renal vasoconstriction, in response to adenosine administration was made half a century ago by Drury & Szent-György (1929), using an isolated perfused kidney preparation. In anesthetized dogs, rats and cats the vasoconstriction in the kidney after intraarterial adenosine administration could be confirmed (Thurau, 1964; Tagawa & Vander, 1970; Osswald, 1975; Osswald et al. 1978 a-c; Spielman & Osswald, 1978 and 1979). Single injections of adenosine into the renal artery induce a rapid fall in renal blood flow which returns to baseline within few seconds. When adenosine (30 - 100 nmoles/min) is infused into the renal artery of dogs renal blood flow is reduced initially but recovers to preinfusion levels within few minutes. Preglomerular afferent arteriolar vasoconstriction in the superficial cortex persists during adenosine infusion (Osswald et al. 1978c; Haas & Osswald, 1981) while vasodilation can be demonstrated in the deep cortex (Spielman et al., 1980).

Adenosine-induced vasoconstriction is somehow coupled to the intrarenal renin-angiotensin system. Sodium diet affects markedly renal vascular response to adenosine: low Na diet accompanied with high plasma renin activity potentiates, while high Na diet with low plasma renin activity attenuates adenosine-induced fall in GFR (Osswald et al., 1978 a,b and 1982). Antagonists of angiotensin II receptors attenuate adenosine-induced vasoconstriction (Spielmann & Osswald, 1979). Angiotensin II and adenosine act in a synergistic manner to lower renal blood flow (Osswald et al., 1978 a) and GFR (Hall et al., 1985). Although the mechanism by which adenosine can induce afferent arteriolar vasoconstriction in the kidney is poorly understood, it is clear that theophylline can antagonize this adenosine action in a fashion that allow to assume a competitive antagonism (Osswald, 1975).

Adenosine derivatives like phenylisopropyladenosine (PIA), 2-chloroadenosine, N-ethylcarboxamide adenosine (NECA), cyclohexyladenosine (CHA) can also induce renal vasoconstriction and a fall in glomerular filtration rate (Murray & Churchill, 1984 and 1985; Pawlowska et al., 1987; Osswald & Beller, 1987). The receptors mediating vasoconstriction could be of the A 1 type since 2-aminophenyladenosine is a very weak and CHA a potent vasoconstricting analog. However, the fact that NECA at low concentration can constrict the renal vasculature (Pawlowska et al., 1987; Osswald & Beller, 1987) does not easily fit into the concept of A 1 receptor mediated vasoconstriction. Binding of ^{3}H-PIA to membranes of dog and rabbit glomeruli reveals a single class of binding sites (Kreutz, 1986; Freissmuth et al., 1987). Since experiments with adenosine receptor antagonists that are subclass specific (A 1/A 2) for the renal vasculature are not yet communicated it is to early to conclude that adenosine-induced vasoconstriction is mediated via A 1 receptors.

There are no investigations known that examine the effects of adenosine on the hemodynamics of the renal medulla. Since the medulla is a low flow tissue in contrast to the cortex one can predict that adenosine might not induce vasoconstriction but rather vasodilation in the vasa recta.

EFFECT ON RENIN SECRETION

Adenosine is a very potent inhibitor of the renal renin secretion rate in anesthetized dogs and rats (Tagawa & Vander, 1970; Osswald et al, 1978 b; Spielman, 1984). This effect is independent from the hemodynamic action of adenosine (Arend et al., 1984) and can be demonstrated in vitro in kidney slices (Churchill & Churchill, 1985) and microdissected afferent arteriols of the glomeruli (Itoh et al., 1985). Rossi & coworkers (1987) could show that the CHA-mediated fall in renin secretory rate of rat renal cortical slices was blocked by pretreating the rats with pertussis toxin. Since theophylline can antagonize the adenosine effects on the renin secretion (Spielman, 1984) the observation of the pertussis toxin blockade suggests that the adenosine receptor on the renin containing cells have very similar coupling mechanisms as the other adenosine receptors in the neuronal tissue or the heart. It appears that micromolar concentrations of NECA can also stimulate renin secretion (Murray & Churchill, 1984 and 1985). Whether this finding has physiological importance awaits further elucidation.

EFFECTS ON TUBULAR EPITHELIUM

There are only few experiments published that investigate the direct effects of adenosine on the tubular epithelium. Infusion of adenosine into the renal artery of dogs results in a fall of urine flow and sodium excretion (Tagawa & Vander, 1970; Osswald, 1975). The reduction in water and electrolyte excretion is the direct result of the reduced glomerular filtration rate. However, small changes in tubular function might be masked by the dominant glomerular action of adenosine.
In a recent publication Seideman et al. (1987) reported changes in free water clearance induced by indomethacin and dipyridamole which was antagonized by theophylline. The suggestion of a possible role of adenosine on the water permeability of the collecting ducts can be supported by earlier findings by Woodcock et al. (1984) who found that NECA stimulates the adenylate cyclase of rat renal papillae homogenates in a concentration dependent manner. The adenosine receptor antagonists, 8-phenyltheophylline and 1,3-diethyl-8-phenylxanthin shifted the concentration response curve to the right. A more detailed study by Arend et al. (1987) revealed that cultured rabbit renal cortical collecting tubules respond with inhibition of cAMP production at low concentrations of CHA, PIA and NECA while higher (micromolar) concentrations of NECA, PIA, CHA lead to an increase in cAMP production of the defined cells. Pertussis toxin pretreatment abolished inhibitory action of adenosine analogues on cAMP production and potentiated the stimulation of cAMP accumulation suggesting that the renal cortical collecting tubules have cell surface receptors for adenosine that are coupled via G proteins to the adenylate cyclase. A support for a physiological role of adenosine in regulating water permeability in the kidney is the finding that 50 nM CHA reduced by 50 % the vasopressin-mediated 5fold increase in cAMP tissue content (Arend et al., 1987). This conclusion is further supported by the observation that adenosine analogs inhibited in isolated perfused rabbit collecting tubules the vasopressin stimulated water movement (Dillingham & Anderson, 1985).

In a number of other epithelia adenosine and its analogs modify electrolyte transport mainly via changes in adenylate cyclase activity: in cultured A 6 amphibian kidney cells (Lang et al., 1985), in the salt gland of the spiny dog fish, Squalus acanthias (Forrest et al., 1982), in the frog cornea (Spinowitz & Zadunaisky, 1979; Osswald et al., 1985), in mammalian ileum and colon (Dobbins et al., 1984). Since cAMP is involved in the control of transepethelial electrolyte transport in many organs it is likely that an increasing number of adenosine effects in these tissues will be observed in the future.

EFFECTS ON RENAL NERVE ENDINGS

The kidney has a rich adrenergic innervation. Renal nerve stimulation or administration of catecholamines modify renal hemodynamics, renin secretion and tubular function (see for ref. Osswald & Greven, 1981). Adenosine is a powerful inhibitor of the release of several neurotransmitters including noradrenaline. This effects appears to be mediated via adenosine A 1 receptors coupled to a G-protein (see for ref. Fredholm et al., 1987). There are only two reports on transmitter release in the kidney modified by adenosine. Hedqvist & Fredholm (1976) and Hedqvist et al. (1978) found that adenosine inhibits noradrenaline release from isolated perfused rabbit kidneys while the effect of exogeneous noradrenaline on vascular resistance was potentiates. These adenosine effects could be antagonized by low micromolar concentrations of theophylline. Thus, the adenosine action on nerve endings in the kidney is similar as in other organs. These findings emphasize the importance of intrarenal adenosine since the contribution of the renal nerves in the control of water and electrolyte balance of the body is substantial.

CONCLUSION

The actions of adenosine on renal function have as in other organs a physiological basis which provides a homeostatic control system depending on the balance of the metabolic rate in the tissue. There are four major structures in the kidney that are modified by adenosine, renal vasculature, renin containing cells, tubular epithelium and nerve endings. Adenosine-mediated fall in glomerular filtration rate, renin secretion and noradrenaline release reduce directly or indirectly the work load of the kidney. The role of adenosine in the direct control of tubular function and medullary hemodynamics remains to be elucidated.

ACKNOWLEDGEMENT

The author acknowledges the secreterial assistance of Ms. R. Fink.

REFERENCES

Arend, L.J., Haramati, A., Thompson, C.I., Spielman, W.S., 1984 Adenosine-induced decrease in renin release: dissociation from hemodynamic effects. Am. J. Physiol., 247: F447-452

Arend, L.J., Sonnenburg, W.K., Smith, W.L. & Spielmann, W.S., 1987 A_1 and A_2 adenosine receptors in rabbit cortical collecting tubule cells. J. Clin. Invest., 79: 710-714

Churchill, P.C. & Churchill, M.C., 1985 A_1 and A_2 Adenosine receptor activation inhibits and stimulates renin secretion of rat renal cortical slices. J. Pharmacol. Exp. Ther., 232: 589-594

Dillingham, M.A. & Anderson, R.J., 1985 Purinergic regulation of basal and arginine vasopressin-stimulated hydraulic conductivity in rabbit cortical collecting tubule. J. Membr. Biol., 88: 277-281

Dobbins, J.W., Laurenson, J.P. & Forrest, J.N., 1984 Adenosine and adenosine analogs stimulate adenosine cyclic 3'5'-monophosphate-dependent chloride secretion in the mammalian ileum. J. Clin. Invest., 74:929-935

Drury, A.N. & Szent-Györgyi, A., 1929 The physiological activity of adenine compounds with special reference to their action upon the mammalian heart. J. Physiol. (London), 68, 213

Freissmuth, M., Hausleithner, V,, Tuisl, E., Nanoff, C. & Schütz, W., 1987 Glomeruli and microvessels of the rabbit kidney contain both A_1- and A_2 -adenosine receptors. Naunyn-Schmiedeberg's Arch. Pharmacol., 335: 438-444

Forrest, J.N., Poeschla, Jr., E. & Rieck, D., 1982 Ribose-specific adenosine receptors mediate cyclic AMP accumulation and chloride transport in the shark rectal gland. Kidney Int., 21: 253 (Abstr.)

Fredholm, B.B., Dunér-Engström, Fastbom, J., Jonzen, B., Lindgren, E., Nordstedt, C., Pedata, F. & van der Ploeg, I., 1987 Interactions Between the Neuromodulator Adenosine and the Classic Transmitters. Topics and Perspectives in Adenosine Research, Springer, Berlin, Heidelberg

Haas, J.A., Osswald, H., 1981 Adenosine induced fall in glomerular capillary pressure: effect of ureteral obstruction and aortic constriction in the Munich-Wistar rat kidney. Naunyn-Schmiedeberg's Arch. Pharmacol., 317: 86-89.

Hall, J.E., Granger, J.P. & Hester, R.L., 1985 Interactions between adenosine and angiotensin II in controlling glomerular filtration. Am. J. Physiol. 248: F340-F346.

Hedqvist, P. & Fredholm, B.B., 1976 Effects of adenosine on adrenergic neurotransmission: prejunctional inhibition and postjunctional enhancement. Naunyn-Schmiedeberg's Arch. Pharmacol., 293: 217-223

Hedqvist, P., Fredholm, B.B. & Ölundh, S., 1978 Antagonistic effects of theophylline and adenosine on adrenergic neuroeffector transmission in the rabbit kidney. Circ. Res., 43: 592-598

Itoh, S., Carretero, O.A. & Murray, R.D., 1985 Possible role of adenosine in the macula densa mechanism of renin release in rabbits. J. Clin. Invest., 76: 1412-1417.

Klahr, S. & Hammermann, M., 1985 Renal metabolism.In: D.W. Seldin and G. Giebisch (Eds.), The Kidney: Physiology and Pathophysiology, Raven Press, New York, pp. 699-718.

Kreutz, R., 1986 Charakterisierung von Adenosinbindungsstellen in der Niere, Inaugural-Dissertation, Universität Bonn.

Lang, M.A., Preston, A.S., Handler, J.S. & Forrest, J.N.Jr., 1985 Adenosine stimulated sodium transport in kidney A 6 epithelia in culture. Am. J. Physiol., 249: C330-C336

Muray, Rd.D. & Churchill, P.C., 1984 Effects of adenosine receptor antagonists in the isolated, perfused kidney. Am. J. Physiol., 247: H343-H348.

Murray, R.D. & Churchill, P.C., 1985 Concentration dependency of the renal vascular and renin secretory responses to adenosine receptor agonists. J. Pharm. Exp. Ther., 232: 189-193.

Osswald, H., 1975 Renal effects of adenosine and their inhibition by theophylline. Naunyn Schmiedeberg's Arch. Pharmacol., 288: 79-86.

Osswald, H., 1983 Adenosine and renal function. In R.M. Berne, T.W. Rall & R. Rubio (Eds.), Regulatory function of Adenosine, M. Nijhoff Publishers, The Hague, Boston, London, pp. 399-415.

Osswald, H., Hermes, H.H. & Nabakowski, G., 1982 The role of adenosine in signal transmission of tubuloglomerular feedback. Kidney Int., 22, Suppl. 12: S136-S142.

Osswald, H., Nabakowski, G. & Hermes, H., 1980 Adenosine as a possible mediator of metabolic control of glomerular filtration rate. Int. J. Biochem., 12: 263-267

Osswald, H., Schmitz, H.-J. & Heidenreich, O., 1978a Adenosine response of the rat kidney after saline loading, sodium restriction and hemorrhagia. Pflüger's Arch., 357: 323-333.

Osswald, H., Schmitz, H.-J. & Kemper, R., 1978b Renal action of adenosine: effect on renin secretion in the rat. Naunyn-Schmiedeberg's Arch. Pharmacol., 303: 95-99.

Osswald, H., Spielman, W.S. & Knox, F.G., 1978c Mechanism of adenosine mediated decrease in glomerular filtration rate. Cir. Res., 43: 465-469.

Osswald, H., Sander, S., Poeschla, E. & Forrest, J.N. 1985 Adenosine-receptors in the salt gland of Squalus acanthias. Antiasthma Xanthines and Adenosine, Eds. K.-D. Andersson & C.G. Persson, Excerpta Medica, Amsterdam, pp. 435-439

Osswald, H. & Greven, J., 1981 Effects of adrenergic activators and inhibitors on kidney function. Handbook of Experimental Pharmacology, L. Szekeres, Springer, Berlin, Heidelberg, New York, 54/2: 243-288

Pawlowska, D., Granger, J.P. & Knox, F.G., 1987 Effects of adenosine infusion into renal interstitium on renal hemodynamics. Am. J. Physiol. 252 (Renal fluid electrolyte physiol.) 21: F678-F682.

Rossi, N.F., Churchill, P.C. & Churchill, M.C., 1987 Pertussis toxin reverses adenosine receptor-mediated inhibition of renin secretion in rat renal cortical slices. Life Sci., 40 (5): 481-487.

Spielman, W.S., Britton, S.L. & Fiksen-Olsen, M.J., 1980 Effect of adenosine on the distribution of renal blood flow in dogs. Cir. Res., 46: 449-456.

Spielman, W.S. & Osswald, H., 1978 Characterization of the postocclusive response of renal blood flow in the cat. Am. J. Physiol., 235: F286-F290.

Spielman, W.S. & Osswald, H., 1979 Blockade of postocclusive renal vasoconstriction by an angiotensin II antagonist: evidence for an angiotensin-adenosine interaction. Am. J. Physiol., 237: F463-F467.

Spinowitz & Zadunaisky, 1979 Action of adenosine on active chloride transport of isolated frog cornea. Amer. J. Physiol., 237: F121-127

Tagawa, H. & Vander, A.G., 1970 Effects of adenosine compounds on renal function and renin secretion in dog. Cir. Res., 26: 327-338.

Thurau, K., 1964 Renal hemodynamics, Am. J. Med. 36, 689-719.

Woodcock, E., Loxley, R., Leung, E. & Johnston, C.I., 1984 Demonstration of R_A-Adenosine receptors in the rat renal papillae. Biochem. Biophys. Res. Com., 121: 434-440

RELEASE AND IN VIVO ACTIONS OF ADENOSINE AND ATP

DEMONSTRATION OF RELEASE OF ATP FROM CENTRAL AND PERIPHERAL NERVES

THOMAS D. WHITE

Department of Pharmacology, Dalhousie University, Halifax, Nova Scotia, Canada B3H 4H7.

INTRODUCTION

There has been considerable resistance to the concept that ATP can, under certain circumstances, be released from cells to exert extracellular actions. However, there is mounting evidence that ATP exerts significant extracellular effects in a wide variety of tissues and organs. ATP can act at extracellular, excitatory P_{2X}-purinoceptors to increase Na^+ and K^+ conductance and consequently depolarize cells. ATP can also activate inhibitory P_{2Y}-purinoceptors to open apamin-sensitive K^+-channels and hyperpolarize cells. Extracellular ATP can act as a substrate for ecto-ATPases (see Pearson, 1985, for review); by analogy to other ATPases, it is tempting to speculate that ecto-ATPase performs some essential cellular function but this remains to be established. Ehrlich et al. (1986) have demonstrated the presence of an ecto-protein kinase which utilizes extracellular ATP for the phosphorylation of specific membrane proteins. Finally, extracellular ATP can be degraded by ectonucleotidases to form adenosine (Pearson, 1985) which can in turn act at inhibitory P_1-purinoceptors in a variety of tissues.

In order to produce these extracellular effects, ATP must first be released from cells. The demonstration of ATP release is made particularly difficult because of its rapid extracellular metabolism primarily to adenosine (see White, 1984, 1985a,b, in press, for reviews). In this chapter, I will review some studies which have demonstrated the release of ATP its possible functions.

NEURONAL RELEASE OF ATP

Sources of ATP for neuronal release. ATP is stored

with acetylcholine within cholinergic synaptic vesicles (Dowdall et al., 1974) and with noradrenaline within sympathetic vesicles (Lagercrantz & Stjarne, 1974; Fried et al., 1978). It is also possible that ATP may exist as the sole neurotransmitter in so-called purinergic nerves (Burnstock, 1979, 1981, 1986), although the evidence for this is not conclusive (White, in press).

Central nervous system. Application of exogenous ATP can produce excitatory effects in the CNS (Phillis & Wu, 1981). Direct electrical stimulation of rat sensory motor cortex results in a 30-fold increase in the amounts of endogenous ATP released into cortical cups (Wu & Phillis, 1978). In addition, depolarization of rat brain synaptosomes by elevated extracellular K^+ or veratridine releases ATP (White, 1977, 1978). K^+-evoked release of ATP is Ca^{2+}-dependent and resembles the secretion of neurotransmitters. However, the veratridine-evoked release of ATP is increased rather than decreased in the absence of Ca^{2+}, a situation also observed for the veratridine-evoked release of GABA from brain slices (Szerb, 1979).

The relative regional distribution for K^+-evoked synaptosomal release of ATP is striatum > cortex > medulla > hypothalamus > cerebellum, whereas veratridine-evoked release is medulla > striatum > hypothalamus > cortex > cerebellum (Potter & White, 1980). Although White et al. (1980) showed that botulinum toxin type A inhibits the release of [^{14}C]acetylcholine but not ATP from rat brain synaptosomes suggesting that ATP release does not occur primarily from cholinergic synaptosomes, Richardson & Brown (1987) have recently demonstrated K^+- and veratridine-evoked release of ATP from affinity purified cholinergic nerve terminals isolated from rat caudate nucleus. The ratio of released acetycholine to ATP was 9:1, similar to the ratio found in cholinergic vesicles. It is possible that the release of ATP from cholinergic synaptosomes observed by White et al. (1980) was masked by the release of ATP from other types of terminals. However, release of ATP does not appear to arise primarily from catecholaminergic synaptosomes since destruction of these nerves by pretreatment with intraventricular 6-hydroxydopamine does not substantially diminish K^+-evoked release of ATP (Potter & White, 1982).

Spinal cord. There is evidence that ATP might be a mediator of excitatory sensory inputs to the spinal cord. Application of ATP excites normally quiescent neurons in the spinal cord (Phillis & Wu, 1979; Jahr & Jessel, 1983;

Fyffe & Perl, 1984; Salter & Henry, 1985). Although it has been been suggested that ATP may be a neurotransmitter mediating nociceptive sensory inputs to the spinal cord (Jahr & Jessel, 1983), more recent studies suggest that ATP may mediate low threshold sensory inputs such as tactile mechanoreception (Fyffe & Perl, 1984; Salter & Henry, 1985). In contrast to ATP, adenosine and its anologues elicit antinociceptive effects when administered intrathecally and it appears that the spinal antinociceptive effects of morphine may be mediated indirectly by the release of adenosine (Sawynok et al., 1986; Delander & Hopkins, 1986; Sweeney et al., in press).

ATP appears to be released during sensory nerve stimulation. Antidromic stimulation of the great auricular nerve of the rabbit ear releases ATP into venous perfusates and produces vasodilation (Holton & Holton, 1954; Holton, 1959). It is likely that ATP is also released in the dorsal horn of the spinal cord. K^+ and veratridine release 2-3 times more ATP from dorsal than from ventral spinal cord synaptosomes but only the K^+-evoked release is Ca^{2+}-dependent (White et al., 1985). This differential release of ATP from dorsal vs ventral cord synaptosomes is consistent with a possible neurotransmitter function for ATP in sensory pathways. However, unilateral section of the sciatic nerve, which depletes the substance P content of the ipsilateral dorsal horn (Jessel et al., 1979), does not diminish subsequent evoked release of ATP from dorsal cord synaptosomes (White et al., 1985). It is possible that this treatment does not deplete the releasable stores of ATP from afferent terminals in the dorsal spinal cord.

Motor nerves. ATP potentiates the action of acetylcholine on the rat hemidiaphragm (Ewald, 1976). In addition, ATP triggers an influx of monovalent cations into cultured chick myotubes (Kolb & Wakelam, 1983; Haggblad et al., 1985; Hume & Honig, 1986). These results are consistent with a possible cotransmitter role for ATP with acetylcholine in motor nerves. However, Hume & Honig (1986) were unable to demonstrate a noncholinergic postsynaptic component to motor nerve stimulation which might be attributed to ATP. ATP also exerts presynaptic effects. ATP modulates Ca^{2+} uptake and inhibits the release of acetylcholine at the neuromuscular junction (Silinsky & Ginsborg, 1983; Lindgren & Smith, 1987). These effects of ATP are not mediated by the activation of P_1-purinoceptors.

Release of ATP has been demonstrated during stimulation

of the phrenic nerve innervating the rat hemidiaphragm (Silinsky, 1975). ATP release persists in the presence of tubocurarine and is not evoked by high doses of carbachol, indicating that it is presynaptic in origin. Further evidence supporting the concept that ATP is released presynaptically from motor nerve terminals has been obtained with cholinergic synaptosomes prepared from the electric organ of the **Torpedo**, where depolarization releases ATP (Morel & Meunier, 1981).

Digestive System. There is evidence that ATP may be a nonadrenergic, noncholinergic (NANC) neurotransmitter in the smooth muscle of certain portions of the gut (see White, in press, for review). For instance, ATP mimicks the relaxant responses to NANC nerve stimulation in the guinea-pig taenia coli (Cocks & Burnstock, 1979) and duodenum (Manzini et al., 1985, 1986). On the other hand, it is unlikely that ATP is the NANC neurotransmitter in the stomach (De Beurme and Lefebvre, 1987) or ileum (Bauer & Kuriyama, 1982; Matusak & Bauer, 1986).

[^{3}H]Adenosine is incorporated into [^{3}H]ATP in the taenia coli and [^{3}H]purines are subsequently released during transmural electrical stimulation (Su et al., 1971; Rutherford & Burnstock, 1978). Virtually all of the released radioactivity is detected as [^{3}H]nucleosides so that one cannot determine whether [^{3}H]ATP or [^{3}H]nucleoside is released. Only about 33% of the released [^{3}H]purine appears to arise from the nerves, the remainder occurring from the muscle. Burnstock et al. (1978b) reported TTX-sensitive release of endogenous ATP during transmural electrical stimulation of taenia coli, but a more recent attempt was not successful (White et al., 1981).

The release of endogenous ATP which, following metabolism by ecto-nucleotidases, could provide a source of extracellular adenosine to act at P_1-purinoceptors, has been detected from nerve varicosities isolated from the myenteric plexus of guinea-pig ileum (White & Leslie, 1982). Ca^{2+}-dependent release of ATP is observed following depolarization with K^+ or veratridine, or exposure to acetylcholine or 5-HT (White & Leslie, 1982; White, 1982; White & Al-Humayyd, 1983; Al-Hummayd & White, 1985a,b). Chemical sympathectomy with 6-hydroxydopamine greatly reduces the evoked release of ATP, suggesting that much of the release occurs from noradrenergic varicosites (White & Al-Humayyd, 1983; Al-Humayyd & White, 1985a,b). However, the α_2-adrenoceptor agonist, clonidine, inhibits the release of [^{3}H]noradrenaline, but not the release of

ATP, from isolated myenteric varicosities (Hammond, MacDonald & White, submitted for publication), an observation which appears to be inconsistent with the concept that ATP and noradrenaline are co-secreted from the same synaptic vesicles. It is possible that ATP and noradrenaline are released from different populations of varicosities in the guinea-pig ileum both of which are susceptible to destruction by 6-hydroxydopamine; however, 6-hydroxydopamine appears to be selective in its toxic effects insofar as serotonergic varicosities are not destroyed (Al-Humayyd & White, 1985b). It is also possible that ATP is released from noradrenergic varicosities but from a different intracellular site which is not subject to modulation by α_2-adrenoceptors. Finally, [^{3}H]noradrenaline may not label the pool of endogenous noradrenaline homogeneously and only the radiolabelled pool may be subject to modulation by α_2-adrenoceptors.

Release of endogenous ATP has been detected when rat anococcygeus muscle is stimulated transmurally (Burnstock et al., 1978a). Exogenous ATP produces e.j.p.'s and contracts the anococcygeus (Stone, 1983; Byrne & Large, 1984), suggesting that this nucleotide could be an excitatory neurotransmitter in this tissue.

Finally, there is evidence that ATP may be the NANC inhibitory neurotransmitter in the duodenum (Manzini et al., 1985,1986) and internal anal sphincter (Lim & Muir, 1986), and an excitatory neurotransmitter in the colon and rectum (Hedlund et al., 1986). ATP may also mediate cAMP-independent glycogenolysis in hepatocytes (Keppens & De Wulff, 1985) and exogenous ATP produces relaxations in the gallbladder (Davison et al., 1978) and promotes secretion in the pancreas (Chapal and Loubatieres-Mariani, 1981) and the parotid gland (Gallacher, 1982). With the exception of the gallbladder, release of ATP has not yet been demonstrated in any of the above situations. Recently, Takahashi et al.(1987) have demonstrated the release of endogenous ATP when superfused strips of guinea-pig gallbladder were stimulated transmurally. However, TTX reduced the contractile response by 97% but only diminished ATP release by 20%. The latter observation is not consistent with ATP being released as a result of propagated action potentials in nerves.

Bladder detrusor. There is evidence, although not conclusive, supporting the possibility that ATP may be an excitatory neurotransmitter in the bladder detrusor (see White, in press, for review). Burnstock et al. (1978b) detected the release of ATP when guinea-pig bladder strips were stimulated transmurally. Release was TTX-sensitive, suggesting that it occurred as the result of propagated action potentials in nerves. In contrast, Chaudhry et al. (1984) were unable to detect the release of ATP from strips of rabbit detrusor muscle using very similar conditions to those of Burnstock et al. (1978b). Although these results do not appear to support the hypothesis that ATP is released as a neurotransmitter in the rabbit bladder, it is possible that ATP was released but that it was rapidly degraded extracellularly to levels below the limits of detection.

Vas deferens. There is excellent evidence that ATP is released from sympathetic nerves to generate e.j.p.'s and the rapid twitch response in the vas deferens (see White, in press, for review). Westfall et al. (1978) detected the release of [^{3}H]purines, primarily [^{3}H]adenosine, following transmural electrical stimulation of guinea-pig vas deferens preloaded with [^{3}H]adenosine. However, it was not certain whether these [^{3}H]purines reflected released [^{3}H]ATP or other purines and whether release occurred pre- or postsynaptically. White et al. (1981) detected the release of endogenous ATP from superfused vas deferens but release was not TTX-sensitive indicating that it was not the result of propagated action potentials in nerves. Recently, we have succeeded in demonstrating TTX-sensitive release of endogenous ATP when the hypogastric nerve and only a small portion of the smooth muscle are stimulated with a suction electrode placed on the prostatic end of the guinea-pig vas deferens (Lew & White, 1987). Contraction of the muscle with phenylephrine does not release ATP, indicating that release is not due to contraction of the muscle **per se**. Release persists when postsynaptic responses are blocked by prazosin and α,β-methylene ATP, indicating that ATP originates presynaptically from the sympathetic nerve varicosities. Kirkpatrick & Burnstock (1987) have also recently demonstrated the release of endogenous ATP during transmural stimulation of the superfused vas deferens. 6-Hydroxydopamine and guanethidine diminished release, indicating that release occurred from sympathetic nerves. Both the release of ATP and the rapid twitch response were unaffected by pretreatment with reserpine, which is consistent with the possibility that reserpine selectively depletes noradrenaline, but not ATP, from sympathetic vesicles. These demonstrations of ATP release provide important support to the hypothesis that ATP functions as an excitatory neurotransmitter in the vas deferens.

Blood vessels. There is mounting evidence supporting the possibility that ATP is an excitatory cotransmitter with noradrenaline in sympathetic nerves innervating blood vessels (Sneddon & Burnstock, 1984; Ishikawa, 1985; Kugelgen & Starke, 1985; Suzuki, 1985; Muramatsu, 1986; Vidal et al., 1986; reviewed by White, in press). Numerous studies have demonstrated the release of radiolabelled purines during stimulation of sympathetic nerves in blood vessels (see Su, 1983, for review). Most of the radiolabel appears as adenosine rather than ATP, presumably because the released ATP is rapidly degraded to adenosine by ecto-nucleotidases. However, it is possible that adenosine is itself released both pre- and postsynaptically during the stimulation of sympathetic nerves in blood vessels. A preliminary report describing the corelease of endogenous ATP with noradrenaline from sympathetic nerves of the rat caudal artery suggests that ATP may account for 70% of the total purine released (Sedaa et al., 1986). It is essential that the release of endogenous ATP from sympathetic nerves be established unequivocally in order to substantiate the proposition that ATP functions as an excitatory cotransmitter with noradrenaline in blood vessels.

Sympathetic neurons in culture. Depolarization of rat sympathetic neurons cultured on islands of heart cells results in NANC hyperpolarization and inhibition of beating of the heart cells (Furshpan et al, 1986). These responses are antagonized by 8-phenyltheophylline or adenosine deaminase, suggesting that they are mediated by adenosine acting at P_1-purinoceptors. Wolinsky & Patterson (1985) have shown that depolarization of cultured sympathetic neurons releases [^{3}H]purines, including nucleotides. Although very little [^{3}H]ATP is detected, the authors state that endogenous ATP is also released. These findings support the concept that ATP may be released from cultured sympathetic neurons where it is metabolized extracellularly to adenosine; the latter could then act at P_1-purinoceptors to produce inhibitory effects on cardiac cells.

Ganglia. Application of exogenous ATP depolarizes frog sympathetic ganglia in culture (Siggins et al., 1977; Akasu et al., 1981,1983; Akasu & Koketsu, 1985). ATP also acts presynaptically to inhibit the release of acetycholine (Silinsky & Ginsborg, 1983). Release of ATP from ganglia has not been studied extensively. Kato et al. (1974) were unable to detect the release of ATP from cat superior cervical ganglia during preganglionic nerve stimulation or depolarization. However, it is possible that the ATP was rapidly degraded extracellularly to levels below the limits of detection of their assay system.

CONCLUSIONS

Release of ATP has been demonstrated from a variety of central and peripheral nerves. However, in many cases where extracellular ATP is thought to perform important functions (most notably as a neurotransmitter in sympathetic nerves innervating blood vessels), the unequivocal release of endogenous ATP has yet to be demonstrated. Studies of ATP release are essential to confirm potential sources of extracellular ATP to perform these functions.

REFERENCES

Akasu, T., Hirai, K. & Koketsu, K., 1981, Br. J. Pharmac., **74**, 505.

Akasu, T., Hirai, K. & Koketsu, K., 1983, Brain Res., **258**, 313.

Akasu, T. & Koketsu, K., 1985, Br. J. Pharmac., **84**, 525.

Al-Humayyd, M. & White, T.D., 1985a, Br. J. Pharmac., **84**, 27.

Al-Humayyd, M. & White, T.D., 1985b, J. Pharmac. exp. Ther., **233**, 796.

Bauer, V. & Kuriyama, H., 1982, J. Physiol. Lond., **332**, 375.

Burnstock, G., 1979 Past and current evidence for the purinergic nerve hypothesis. In Physiological and Regulatory Functions of Adenosine and Adenine Nucleotides, edited by H.P. Baer & G.I. Drummond (Raven Press) p.3.

Burnstock, G., 1981, J. Physiol. Lond., **313**, 1.

Burnstock, G., 1986, Acta physiol. scand., **126**, 67.

Burnstock, G., Cocks, T. & Crowe, R., 1978a, Br. J. Pharmac., **64**, 13.

Burnstock, G., Cocks, T., Kasakov, L. & Wong, H., 1978b, Eur. J. Pharmacol., **49**, 145.

Byrne, N.G. & Large, W.A., 1984, Br. J. Pharmac., **83**, 751.

Chapal, J. & Loubatieres-Mariani, M.M., 1981, Br. J. Pharmac., **73**, 105.

Chaudhry, A., Downie, J.W. & White, T.D., 1984, Can. J. Physiol. Pharmacol., **62**, 153.

Cocks, T. & Burnstock, G., 1979, Eur. J. Pharmacol., **54**, 251.

Davison, J.S., Al-Hassani, M., Crowe, R. & Burnstock, G., 1978, Pflugers Arch., **377**, 43.
De Beurme, F.A. & Lefebvre, R.A., 1987, Br. J. Pharmac., **91**, 171.
Delander, G.E. & Hopkins, C.J., 1986, J. Pharmac. exp. Ther., **239**, 88.
Dowdall, M.J., Boyne, A.F. & Whittaker, V.P., 1974, Biochem. J., **140**, 1.
Ehrlich, Y.H., Davis, T.B., Bock, E., Kornecki, E. & Lenox, R.H., 1986, Nature, **320**, 67.
Ewald, D.A., 1976, J. Membrane Biol., **29**, 47.
Fredholm, B.B. & Sollevi, A., 1985, Clin. Physiol., **6**, 1.
Fried, G., Lagercrantz, H. & Hokfelt, T., 1978, Neuroscience, **3**, 1271.
Furshpan, E.J., Potter, D.D. & Matsumoto, S.G., 1986, J. Neurosci., **6**, 1099.
Fyffe, R.E.W. & Perl, E.R., 1984, Proc. natn. Acad. Sci. U.S.A., **81**, 6890.
Gallacher, D.V., 1982, Nature, **296**, 83.
Haggblad, J., Eriksson, H. & Heilbronn, E., 1985, Acta physiol. scand., **125**, 389.
Hedlund, H., Fandriks, L., Delbro, D. & Fasth, S., 1986, Acta physiol. scand., **127**, 425.
Holton, F.A. & Holton, P., 1954, J. Physiol. Lond., **126**, 124.
Holton, P., 1959, J. Physiol. Lond., **145**, 494.
Hume, R.I. & Honig, M.G., 1986, J. Neurosci., **6**, 681.
Ishikawa, S., 1985, Br. J. Pharmac., **86**, 777.
Jahr, C.E. & Jessel, T.M., 1983, Nature, **304**, 730.
Jessel, T., Tsunoo, A., Kanazawa, I. & Otsuka, M., 1979, Brain Res., **168**, 247.
Kato, A.C., Katz, H.S. & Collier, B., 1974, Nature, **249**, 576.
Keppens, S. & De Wulff, H., 1985, Biochem. J., **231**, 797.
Kirkpatrick, K. & Burnstock, G., 1987, Eur. J. Pharmacol., **138**, 207.
Kolb, H.-A. & Wakelam, M.J.O., 1983, Nature, **303**, 621.
Kugelgen, I.v. & Starke, K., 1985, J. Physiol. Lond., **367**, 435.
Lagercrantz, H. & Stjarne, L., 1974, Nature, **249**, 843.
Lew, M.J. & White, T.D., 1987, Br. J. Pharmac, in press.
Lim, S.P. & Muir, T.C., 1986, Eur. J. Pharmacol., **128**, 17.
Lindgren, C.A. & Smith, D.O., 1987, J. Neurosci., **7**, 1567.
Manzini, S., Maggi, C.A. & Meli, A., 1985, Eur. J. Pharmacol., **113**, 399.
Manzini, S., Maggi, C.A. & Meli, A., 1986, Eur. J. Pharmacol., **123**, 229.

Matusak, O. & Bauer, V., 1986, Eur. J. Pharmacol., **126**, 199.
Morel, N. & Meunier, F.-M., 1981, J. Neurochem., **36**, 1766.
Muramatsu, I., 1986, Br. J. Pharmac., **87**, 478.
Pearson, J.D., 1985 Ectonucleotidases: measurement of activities and use of inhibitors. In Methods in Pharmacology: Methods Used in Adenosine Research, Vol.6, edited by D.M. Paton (Plenum Press) p.83.
Phillis, J.W. & Wu, P.H., 1981, Prog. Neurobiol., **16**, 187.
Potter, P. & White, T.D., 1980, Neuroscience, **5**, 1351.
Potter, P.E. & White, T.D., 1982, Eur. J. Pharmacol., **80**, 143.
Richardson, p.J. & Brown, S.J., 1987, J. Neurochem., **48**, 622.
Rutherford, A. & Burnstock, G., 1978, Eur. J. Pharmacol., **48**, 195.
Salter, M.W. & Henry, J.L., 1985, Neuroscience, **15**, 815.
Sawynok, J., Sweeney, M.I. & White, T.D., 1986, Br. J. Pharmac., **88**, 923.
Sedaa, K., Bjur, R.A. & Westfall, D.P., 1986, Fed. Proc., **45**, 582.
Siggins, G.R., Gruol, D.L., Padjen, A.L. & Forman, D.S., 1977, Nature, **270**, 263.
Silinsky, E.M., 1975, J. Physiol. Lond., **247**, 145.
Silinsky, E.M. & Ginsborg, B.L., 1983, Nature, **305**, 327.
Sneddon, p. & Burnstock, G., 1984, Eur. J. Pharmacol., **106**, 149.
Stone, T.W., 1983, J. Physiol. Lond., **335**, 591.
Su, C., 1983, Ann. Rev. Pharmacol. Toxicol., **23**, 397.
Su, C., Bevan, J.A. & Burnstock, G., 1971, Science, **173**, 336.
Suzuki, H., 1985, J. Physiol. Lond., **359**, 401.
Sweeney, M.I., White, T.D. & Sawynok, J., in press, J. Pharmac. exp. Ther.
Szerb, J.C., 1979, J. Neurochem., **32**, 1565.
Takahashi, T., Kusunoki, M., Ishikawa, Y., Kantoh, M., Yamamura, T. & Utsonomiya, J., 1987, Eur. J. Pharmacol., **134**, 77.
Vidal, M., Hicks, P.E. & Langer, S.Z., 1986, Naunyn-Schmiedebergs Arch. Pharmak., **332**, 384.
Westfall, D.P., Stitzel, R.E. & Rowe, J.N., 1978, Eur. J. Pharmacol., **50**, 27.
White, T.D., 1977, Nature, **267**, 67.
White, T.D., 1978, J. Neurochem., **30**, 329.
White, T.D., 1982, Eur. J. Pharmacol., **79**, 333.
White, T.D., 1984, Prog. Neurobiol. Biol. Psychiat., **8**, 487.

White, T.D., 1985a Release of ATP from central and peripheral nerve terminals. In Purines: Pharmacology and Physiological Roles, edited by T.W. Stone (Macmillan) p.95.

White, T.D., 1985b The demonstration and measurement of adenosine triphosphate release from nerves. In Methods in Pharmacology: Methods Used in Adenosine Research, Vol.6, edited by D.M. Paton (Plenum Press) p.43.

White, T.D., in press, Pharmacol. Ther.

White, T.D. & Al-Humayyd, M., 1983, J. Neurochem., **40**, 1069.

White, T.D., Downie, J.W. & Leslie, R.A., 1985, Brain Res., **334**, 372.

White, T.D. & Leslie, R.A., 1982, J. Neurosci., **2**, 206.

White, T. Potter, P., Moody, C. & Burnstock, G., 1981, Can. J. Physiol. Pharmacol., **59**, 1094.

White, T.D., Potter, P. & Wonnacott, S., 1980, J. Neurochem., **34**, 1109.

Wolinski, E.J. & Patterson, P.H., 1985, J. Neurosci., **7**, 1680.

Wu, P.H. & Phillis, J.W., 1978, Neurochem. Res., **3**, 563.

USE OF ENPROFYLLINE TO DELINEATE ENDOGENOUS ACTIONS OF ADENOSINE

Carl G A Persson

Pharmacological Laboratory
AB Draco
Box 34
S-221 00 Lund
Sweden

Dept. of Clinical Pharmacology
University Hospital
S-221 85 Lund
Sweden

INTRODUCTION

A conventional view of drug discovery is that it is derived from progress in basic research carried out at institutions outside the drug industry. However, the reverse may also be a reasonable description. The discovery of an innovative drug and the pharmacology involved in its development may significantly contribute to progress in basic research. A new drug more often than not emerges because unexpected properties have been revealed with a compound. Hence novel drugs precede and are frequently a source of development of new basic knowledge, in particular in pharmacology (Persson 1982, Maxwell 1984) and in pathophysiology too. In some small measure this is borne out by our own experience with xanthines. The pharmacological effects of these compounds contrasted accepted and evolving theories on cellular modes of drug action in asthma. Our structure-activity work also gave us tools such as enprofylline to evaluate proposed mechanisms of action of xanthines. Whereas enprofylline data do not support a role for adenosine in asthmatic lungs they indirectly support the view that adenosine has a regulatory function in select extrapulmonary tissues.

Xanthines, a group of compounds with complex pharmacology

Xanthines have a variety of potential drug applications ranging from relaxation of smooth muscle and through antiinflammatory effects to ability to stimulate sperm motility. Recently they have been shown to have anticancer effects and effects on blood rheology. The possibility of using xanthines as pesticidal agents has been raised (for references see Persson 1985a). The diverse properties of xanthines presumably reflect a plethora of mechanisms of action most of which may be as yet unknown.

The xanthines are currently used clinically mainly as psychotropic stimulants (caffeine) and as antiasthma drugs (theophylline). The therapeutic and safety limits of plasma and tissue concentrations of these xanthines severely restrict the number of mechanisms of action which might be clinically significant. Adenosine antagonism is one such potent mechanism.

Xanthines in asthma

We owe much of our knowledge of asthma and its treatment to Dr Salter (Salter 1860, Persson 1985). One of his achievements was to select large cups of strong coffee as a primary treatment of this disease. He made this choice from the extremely complex therapeutic armamentarium available to the medical profession in the 19th century. He was careful with the dose regimen and, as a result, had great success with his caffeine remedy in many asthma patients including himself.

Interestingly, Salter produced several arguments in favour of the view that the usefulness of caffeine in this disease was related to its "CNS-exaltatory" action. Excitatory extra-pulmonary effects of xanthines are again attracting interest and improved diaphragmatic contractility is advocated as being an important therapeutic action of xanthines in asthma. The xanthine compound theophylline is presently dominating the oral drug treatment of asthma worldwide. Theophylline is almost identical with caffeine, only slightly more potent.

During the last decade there has been an increasing emphasis on pulmonary antiinflammatory effects of xanthines. Such effects may well be therapeutically important. They involve a number of cells other than bronchial smooth muscle. The lack of precise knowledge of pulmonary (and extrapulmonary) target cells makes it difficult to evaluate critically the mechanism(s) of action. Examination of a role for adenosine in the anti-asthma actions of xanthines may involve airway epithelial, endothelial, and smooth muscle cells as well as a variety of inflammatory cells that accumulate in asthmatic airways. Furthermore, adenosine mechanisms involved in the regulation of skeletal muscle contractility would have to be examined if improved diaphragmatic and intercostal muscle action should turn out to be an important property of antiasthma xantines.

Some actions of adenosine and theophylline, in particular on mastcells and smooth muscle, have led investigators to propose that adenosine antagonism, rather than phosphodiesterase inhibition, may explain the antiasthma efficacy of xanthines (Aviado and Salem 1976, Marquardt et al 1978, Fredholm et al 1979, Welton 1980, Snyder 1981, Holgate et al 1984). This view has been widely disseminated and work has been initiated to find novel adenosine-antagonists for the treatment of asthma.

It is against the above background that the novel xanthine enprofylline should be viewed. Enprofylline is a poor phosphodiesterase inhibitor and, within reasonable concentrations, it lacks the ability to antagonize pharmacological effects of adenosine. Enprofylline is without CNS excitatory effects and does not increase diaphragm-contractility. However, it is effective in asthma (see Andersson and Persson 1985, Persson 1986).

Phases in the development of enprofylline

The work carried out in my group during the last decade and a half with xanthines may be summarized in a few sentences. We started to look for antiasthma drugs among a variety of old and novel phosphodiesterase inhibitory compounds. This was soon followed by a more controversial interest in xanthine derivatives. (These 'ancient' drugs were then, at the beginning of 1970s, considered obsolete, at least by most opinion leaders outside North America.) However, in simple test systems, such as guinea-pig and human airways in vitro and histamine-challenged conscious guinea-pigs, these substances compared well with potent experimental phosphodiesterase inhibitory compounds, with prostaglandins, Ca^{++}-antagonists, and several single mediator antagonists. We are still getting data on bronchodilation and airway antiinflammatory actions suggesting that xanthines are competitive antiasthma research compounds.

Proposed mechanisms of action of xanthines such as cAMP- or cGMP-phosphodiesterase inhibition, antagonism of prostaglandins, adrenaline release, and adenosine antagonism appeared to us to be little helpful in the search for a novel antiasthma drug. We were rather guided by the fact that xanthine derivatives exhibited a great variety of pulmonary and extrapulmonary actions. Frequent pharmacological discoveries transformed the "boring" drug screening work, into an exciting search for unexpected properties. After collecting structure-activity data of xanthines for several years we had a diversity of action profiles to select from. The actions and the toxicology of 3-propylxanthine, enprofylline, were particularly promising (Persson and Kjellin 1981). Once the pulmonary and extrapulmonary effects of enprofylline had been roughly established we had to look for explanations of its particular pharmacological profile. We then again took up studies of adenosine antagonism.

During the last seven years an escalating number of clinical studies has been performed confirming and extending the pulmonary effect similarities and extrapulmonary effect differences between enprofylline and theophylline. The differences may suggest physiological roles for adenosine in man.

Structure-activity of xanthines

The xanthine molecule can be substituted at at least four different positions to produce xanthine derivatives. The relationships between structure and activity are complex. A few points related to antiasthma xanthines may be mentioned. Substitutions at the 7- (or 9-) position result in low potency, whereas substitutions at 1- and 8-positions may be associated with increased potency, however, at the expense of an out of proportion increased toxicity. Substitutions at 1- and 8-positions are also critical for potent adenosine antagonism. The 8-position may also be used to produce compounds which selectively inhibit cGMP phosphodiesterase. By proper substitution at the 3-position antiasthma potency increases. Furthermore, compounds having an alkyl only in the 3-position may lack adenosine antagonism and have relatively few pharmacological actions. By increasing the bulkiness in this position airway potency increases with an optimum for enprofylline 3-propyl xanthine. Enprofylline is also more potent than theophylline (1,3-dimethyl xanthine).

Effect similarities between enprofylline and theophylline would reflect a shared mode of action that is independent of adenosine

In contrast to theophylline enprofylline is a poor adenosine antagonist without significant ability to antagonize a large variety of pharmacological actions of adenosine. This has been shown in animal and human tissues harbouring either or both subtypes of adenosine receptors (Persson 1986a). Enprofylline is a pharmacologically active compound albeit with fewer effects than theophylline. In those actions which are shared there are far-reaching qualitative similarities between enprofylline and theophylline, except for potency. Potentially important airway actions such as attenuation of inflammatory stimulus-induced plasma exudation (Persson 1986b) are produced by both xanthines. Relaxant responses in large and small airway smooth muscle are shared in detail, indicating that theophylline and enprofylline have a common mechanism(s) of action. Ca^{++}-antagonism, prostaglandin antagonism, cAMP-phosphodiesterase inhibition, cGMP-phosphodiesterase inhibition and adrenalin release are proposed mechanisms which in our hands seem non-contributory to the common mode of action (see Andersson and Persson 1985). Our data show either a lack of potency of the xanthines (eg concerning cAMP-phosphodiesterase inhibition), a qualitative difference between xanthines and prototype mechanism-utilizing compounds (eg Ca^{++}-antagonism), or inconsistent structure-activity relationships between rank order of bronchodilation and effects on the mechanism (eg cGMP phosphodiesterase inhibition). It is, therefore, difficult to point out any action on receptors, enzymes or other subcellular mechanism that would explain those effects which are exerted by both enprofylline and theophylline.

In asthmatic subjects enprofylline (2-5 µg/ml plasma) and theophylline (5-20 µg/ml) have comparable efficacies when given as prophylactic or sympatomatic drugs in acute and long term treatments. The only exception may be when bronchoconstriction is acutely provoked by letting asthmatic subjects inhale adenosine. Theophylline is then relatively more effective than enprofylline (unpublished work by Holgate et al). Taken together the available documentation on antiasthma efficacy of enprofylline suggests that adenosine antagonism in the airways or elsewhere may not be important for the therapeutic effect of xanthines in asthma. As a corollary it follows that extrapulmonary effects may be non-contributory to antiasthma effects of xanthines. Furthermore, the similarities in antiasthma efficacy between enprofylline and theophylline are indirect evidence in favour of the view that adenosine is not a major mediator of asthma.

Both enprofylline and theophylline exert a number of cardiac stimulant actions. It has been proposed that adenosine may have an important role in reducing dromotropic, chronotropic, and inotropic activities of the heart, in particular when the heart is stressed by sympathetic β-adrenoceptor stimulation. Enprofylline, being a poor adenosine antagonist, might then be a safer drug for the asthmatic heart than theophylline (Persson et al 1983). However, animal in vivo work has not differentiated cardiac stimulant effects of enprofylline and theophylline (Persson and Erjefält 1983). Also in asthmatic subjects cardiac actions of the two xanthines are similar, with no tendency for theophylline alone or together with the β_2-adrenoceptor agonist terbutaline to cause disproportionately (compared with enprofylline) large effects on the heart rhythm (Conradson et al 1987). Under the conditions tested so far the effects of enprofylline and theophylline have not confirmed a significant cardiodepressant or cardioprotective role for endogenous adenosine.

Actions of theophylline not shared by enprofylline may show roles of endogenous adenosine

Methods used in adenosine research have undergone impressive developments in recent years (Paton 1985). However, a number of factors may still make it difficult to delineate specific roles for adenosine, in particular in man. (1) Adenosine receptors are ubiquitous and adenosine may well have different actions even in a single cell type. (2) Administration of the nucleoside itself may not always yield valid data due to its rapid breakdown and limited distribution. (3) When adenosine is given intravascularly its potent vascular action will dominate and adequate doses for effects on a variety of tissues may not readily be given by this route. (4) Even when topical application is employed exogenous adenosine may not reach important biophases in the same way as the endogenously released nucleoside would. (5) Adenosine analogues may not have entirely identical actions to adenosine itself and few if any analogues can be used in human experiments. (6) Caffeine and theophylline may be given to man in adenosine antagonistic doses but these blocking drugs lack specificity. In these circumstances enprofylline would have potential value because it may point out such actions of theophylline (and caffeine) where most likely adenosine is not involved.

By its lack of theophylline-like diuretic effects, free fatty acid releasing effects, gastric secretory effects, and CNS-stimulant behavioural effects enprofylline has given support to the view that adenosine may have physiologically important inhibitory effects in these systems. Enprofylline, in addition to its poor adenosine antagonism, less readily than theophylline passes the blood brain barrier. However, lack of CNS-stimulation is seen at doses of enprofylline producing about the same brain levels of this drug in guinea-pigs as obtained by seizure-inducing theophylline doses (Persson and Erjefält 1983). Also, in asthmatic subjects receiving 24 h infusion of enprofylline with resulting plasma levels as high as 42 µg/ml (mean 23 µg/ml, n=7) there was no sign of any CNS-excitatory effect (unpublished work by Vilsvik, Eriksson, Persson). Headache, nausea and tachycardia did occur, however. (Assuming that theophylline is 3-5 times less potent than enprofylline as a bronchodilator these plasma levels of enprofylline would correspond to very high and toxic levels of theophylline).

CONCLUSION

It is suggested that enprofylline is a helpful tool in the pharmacologic dissection of the roles of endogenous adenosine, in particular in man.

REFERENCES

Andersson, K.-E.& Persson, C.G.A. 1985, Antiasthma xanthines and adenosine. Excerpta Medica Amsterdam.

Aviado, D.M. & Salem, H. 1976, Bronchodilator and anti-asthmatic drugs. In Bronchial Asthma Mechanisms and Therapeutics (Eds E.B. Weiss & M.S. Segal) Little Brown Co., Boston, pp. 715-726.

Conradson, T.-B, Edlundh, G., Olofsson, B., Pahlm, O. & Persson, G. 1987, Arrhytmogenicity from combined bronchodilator therapy in patients with obstrubtive lung disease and concomitant ischemic heart disease. Chest 91, 5-9.

Fredholm, B.B., Brodin, K. & Strandberg, K. 1979, On the mechanism of relaxation of tracheal smooth muscle by theophylline and other cyclic nucleotide phosphodiesterase inhibitors. Acta Pharmacol. Toxicol. 45, 336.344.

Holgate, S.T., Mann, J.S. & Cushley, M.J. 1984, Adenosine as a bronchoconstrictor mediator in asthma and its antagonism by methylxanthines. J. Allergy Clin. Immunol. 74, 302-307.

Marquardt, D.L. Parker, C.N. & Sullivan, T.J. 1978, Potentiation of mast-cell mediator release by adenosine. J. Immunol. 120, 871-878.

Maxwell, R.A. 1984, The state of the art of the science of drug discovery - an opinion. Drug Development Research 4, 375-389.

Paton, D.M. 1985 (editor), Methods in pharmacology volume 6. Methods used in adenosine research. Plenum New York.

Persson, C.G.A. 1982, Some β-adrenoceptor functions of therapeutic interest. Acta Obstet. Gynecol. Scand. Suppl. 108, 7-11.

Persson, C.G.A. 1985a, Subdivision of xanthines, In: Antiasthma xanthines and adenosine edited by K.-E. Andersson and C.G.A. Persson Excerpta Medica Amsterdam, 23-40.

Persson, C.G.A. 1985b, On the medical history of xanthines and other remedies for asthma: a tribute to HH Salter. Thorax 40, 881-886.

Persson, C.G.A. 1986a, Development of safer xanthine drugs for treatment of obstructive airways disease. Journal of Allergy and Clinical Immunology 78, 817-824.

Persson, C.G.A. 1986b, Role of plasma exudation in asthmatic airways. Lancet 2, 1126-1129.

Persson, C.G.A. & Erjefält, I. 1983, Seizure activity in animals given enprofylline and theophylline, two xanthines with partly different mechanisms of action. Arch. Int. Pharmacodyn. Ther. 258, 261-274.

Persson. C.G.A., Erjefält, I. & Andersson, K.-E. 1983, Positive inotropic and chronotropic effects and coronary vasodilation in vitro by two antiasthmatic xanthines with different abilities to antagonize adenosine. Journal of Cardiovascular Pharmacology 5, 778-785.

Persson, C.G.A. & Kjellin, G. 1981, Enprofylline, a principally new antiasthmatic xanthine. Acta Phamracologica et Toxicologica 49, 313-316.

Salter, H.H. 1860, On asthma: its pathology and treatment. Churchill, London.

Snyder, S.H. 1981, Adenosine receptors and the actions of methylxanthines. Trends Neuroschi. 4, 242-244.

Welton, A.F. 1980, Regulatory role of adenosine in antigen-induced histamine-release from lung tissue of actively sensitized guinea-pigs. Biochem. Pharmacol. 29, 1085-1092.

THE ACTIONS OF ADENOSINE ON RESPIRATION

J.A. Ribeiro, E.C. Monteiro & D.S. McQueen*

Laboratory of Pharmacology, Gulbenkian Institute of Science, 2781 Oeiras, Portugal and *Department of Pharmacology, University of Edinburgh, Edinburgh EH8 9JZ, U.K..

INTRODUCTION

Adenosine has several actions on the respiratory system. In adults adenosine: 1) stimulates respiratory ventilation in rats (Monteiro & Ribeiro 1987), rabbits, (Buss et al. 1986), and cats (Monteiro & Ribeiro - unpublished observations), an effect abolished after section of the carotid sinus nerves (glomectomized animals); 2) in cats, increases carotid body chemosensory activity (McQueen & Ribeiro 1981); 3) in humans, stimulates respiratory ventilation, an effect which as in rats or rabbits appears to be mediated by carotid body chemoreceptors (Watt & Routledge, 1985; Biaggioni et al. 1986; Maxwell et al. 1986; Watt et al. 1987b), and increases the hypoxic ventilatory response but does not alter the hypercapnic response (Maxwell et al. 1986). Adenosine analogues given i.v. (Eldridge et al. 1984; Monteiro & Ribeiro 1987) i.p. (Wessberg et al. 1985) or intracerebroventricularly (i.c.v). (Eldridge et al. 1984; Wessberg et al. 1985) cause delayed on onset and prolonged inhibition of respiration in both non-, and glomectomized cats and rats. This effect appears to be mediated through direct inhibition of the central nervous system regulation of respiration, and is antagonized by theophylline. In glomectomized cats, adenosine appears to be responsible for the depression of respiration during hypoxia (Milhorn et al. 1984).

In preterms and neonates: In preterm rabbits the adenosine analogue L-phenylisopropyladenosine (L-PIA) i.p. depresses respiration (Hedner et al. 1984). L-PIA i.p. or applied to the fourth ventricle in neonatal rabbits also inhibits respiration, an effect which is more pronounced in younger than in older animals (Lagercrantz et al. 1984; Runold et al. 1986). These effects are antagonized by theophylline. In neonatal rabbits, both hypoxia and i.v. adenosine depress respiration and after aminophylline this inhibitory effect of adenosine on respiration is abolished and converted into respiratory stimulation; aminophylline also abolishes the respiratory depression induced by hypoxia (Watt et al. 1987a). Both the inhibitory and the excitatory respiratory responses to adenosine i.v. are rapid on onset and short-lasting (Watt et al. 1987a).

Other effects of adenosine on the respiratory system: 1) The bronchoconstrictor effect. The i.v. injection of adenosine in rats causes bronchoconstriction, an effect antagonized by theophylline (Pauwells & Van der Straeten 1986). This bronchoconstrictor effect is dependent upon the anaesthetic used: with urethane instead of pentobarbitone the effect is not observed. Inhaled adenosine causes bronchoconstriction in asthmatics (Mann et al. 1986). 2) The diaphragmatic effect. Theophylline *in vitro*, at therapeutic doses, enhances diaphragmatic contractility (Viires et al. 1986) and antagonizes the adenosine receptors of the rat phrenic nerve endings (Ribeiro & Sebastião 1986a) and *in vivo* it has been shown that aminophylline potentiates diaphragmatic force generation at clinically useful doses (Aubier et al. 1981).

In this chapter emphasis will be on results obtained from experiments on cat carotid body chemoreceptor activity and on rat respiration mediated by carotid body chemoreceptors.

RESULTS

1. The effect of adenosine on carotid body chemoreceptor activity in artificially ventilated cats.

The effect of adenosine. In 1981 McQueen & Ribeiro (1981) described that adenosine i.c. causes a rapid and marked increase of spontaneous carotid body chemoreceptor discharge in the cat, the intensity, duration and onset of which is dose-dependent. Infusion of adenosine also evokes an increase in discharge, which persists for the duration of the infusion, and returns to pre-infusion control levels within 30 s of stopping the infusion. These adenosine infusions were not associated with changes in arterial blood pressure (BP). No tachyphylaxis to the action of adenosine was observed. The peak of discharge for all doses (0.37-374 nmol) occurs between 5 and 10 s after the injections. The adenosine response for the highest dose studied (374 nmol i.c.) lasts about 20 to 30 s. Adenosine increases the average discharge and also increases the maximum discharge frequency (see McQueen & Ribeiro, 1981). The overall increase in discharge caused by adenosine results from a sustained increase throughout the response rather than from a sudden transient increase, such as occurs with acetylcholine (ACh) (cf. McQueen, 1977). These effects are not related to changes in BP. There are no changes in arterial blood gas tensions or pH following administration of adenosine.

Specificity of adenosine. The effect is specific for adenosine since adenosine related compounds such as the stable adenosine triphosphate (ATP) analogue, α-β-methylene ATP causes inhibition in ventilation (McQueen & Ribeiro, 1983) and inosine a metabolite of adenosine deamination is inactive. Similar results with inosine have been obtained recently in humans by Reid et al. (1987) where adenosine but not inosine stimulates ventilation.

The adenosine receptor. Adenosine analogues. 5'-N-ethylcarboxamidoadenosine (NECA) i.c. causes a dose related chemoexcitation which was long-lasting, with discharge remaining elevated for several minutes after the injection and gradually returning to control levels 5-10 min after the high doses of NECA. Although less potent than adenosine as chemoreceptor stimulants, L-PIA and D-PIA were both able to increase chemosensory discharge. The comparison between the two isomers showed L-PIA to be slightly more potent, by a factor of two, in stimulating the carotid chemoreceptors. In Table 1 are summarized the ED_{50} values and the relative potencies for NECA, 2-chloroadenosine (CADO) and L-PIA. It is clear that NECA is more potent than CADO and much more potent than L-PIA. This order of potency together with the

Table 1. ED_{25} and ED_{50} values, and the relative potencies of the adenosine analogues on chemoreceptor activity in the cat, and on spontaneous ventilation in the rat.

		NECA	CADO	L-PIA
CATS*	ED_{50} (nmol)	1.7±0.5	33	131±11.7
	Relative potency	77	4	1
RATS**	ED_{25} (nmol)	0.1	1	5.9
	Relative potency	59	5.9	1

Data from: * McQueen & Ribeiro (1983; 1986)
** Monteiro & Ribeiro (1987)

observation that dibutyryl cyclic AMP mimics the excitatory action of adenosine (Ribeiro & McQueen, 1983), are highly suggestive of the involvement of an A_2 subtype of adenosine receptor.

Alkylxanthines. Theophylline infused i.c. decreased the chemoexcitation induced by NECA (McQueen & Ribeiro 1986). Also, the more potent adenosine receptor antagonist, 8-phenyltheophylline (8-PT i.c.) antagonized the increase in chemoreceptor discharge evoked by adenosine (McQueen & Ribeiro 1986).

Dipyridamole. Intravenous infusion of dipyridamole caused a slight increase in spontaneous chemoreceptor discharge during the period of infusion, and shifted to the left the log dose-response curve to adenosine (McQueen & Ribeiro 1986). Intracarotid infusion of dipyridamole caused a more marked increase in spontaneous discharge. During the i.c. infusion of dipyridamole the increase in discharge evoked by injections of adenosine was greatly potentiated (McQueen & Ribeiro 1986).

Influence of adenosine receptor antagonism on responsiveness of chemoreceptors to hypoxia. Ventilating a cat with hypoxic gas mixture (10% O_2: 90% N_2), instead of room air, increased chemosensory discharge. The responsiveness to hypoxia, in terms of both the slope of the line relating discharge to duration of hypoxia and the steady state discharge (= 100%) on ventilating with 10% O_2, was reduced during infusion of 8-PT. These findings taken together with the observation that chemoexcitation evoked by adenosine is substantially reduced during the 8-PT infusions (McQueen & Ribeiro 1986) suggest that endogenous adenosine might be involved in chemosensory excitation induced by hypoxia.

Structure of carotid body where adenosine acts. To determine in which of the carotid body structures adenosine acts, Ribeiro & McQueen (1983) investigated how adenosine influences the chemoreceptor responses evoked by sodium cyanide (NaCN), CO_2-equilibrated Locke solution, ACh and dopamine. It has been suggested that substances such as NaCN act preferentially on carotid body type I cells, and substances such as ACh and dopamine act predominantly on the sensory nerve endings (Eyzaguirre & Fidone 1980). Responses to NaCN and CO_2-equilibrated Locke solution were slighlty and variably reduced after adenosine injections, and the excitatory responses evoked by ACh and the inhibitory responses evoked by dopamine were both increased after adenosine injections. This sensitization by adenosine suggests that this nucleoside might have a direct action on the sensory nerve endings, which are postsynaptic in relation to type I cells. The mechanism triggered by adenosine to cause chemoreceptor activation might occur at pre- and/or post-synaptic levels as it was already discussed in a previous paper (McQueen & Ribeiro 1981). Considering the post-synaptic level, a possibility can be envisaged at the light of recent results obtained with the rat sympathetic ganglia (Henon & McAfee 1983). In this preparation, CADO , besides inhibiting excitatory post-synaptic potentials (EPSPs) probably through a presynaptic effect, i.e. decreasing the amount of transmitter released per impulse as it does in most of the central and peripheral synapses (for a review see e.g. Ribeiro & Sebastião 1986b), also posseses facilitatory effects on synaptic transmission during repetitive firing. In the presence of CADO more EPSPs exceed threshold and the number of action potentials increases markedly. Whether such a mechanism acting at the carotid body sensory nerve endings also applies to the increase in chemoreceptor discharge caused by adenosine needs to be investigated.

2. Effect of adenosine on spontaneous ventilation mediated by carotid body chemoreceptors in the rat.

The effect of adenosine. Adenosine (0.1-100 nmol) injected i.c. before cutting the carotid sinus nerves increases both the amplitude and the frequency (f) of the respiratory movements. The effect started immediately after the end of the injections and lasted for about 30-40 s for the highest dose (100 nmol). Slight decreases in both BP and heart rate (HR) ocurred also. After section of the carotid sinus nerves the excitatory effect of adenosine on ventilation disappeared, no marked changes in BP were observed and the decrease in HR was more evident. No substantial differences were observed in these excitatory effects after section of the vagal nerves (Monteiro & Ribeiro 1987).

The adenosine receptor. The adenosine analogues. The comparison of the excitatory effects of the adenosine analogues, NECA, CADO and L-PIA showed that NECA>CADO>L-PIA (Table 1). This order of potency and the absence of marked stereoselectivity for the PIA isomers (D-PIA/L-PIA = 0.9) suggest, as it happens with the adenosine receptor that mediates adenosine excitation of cat carotid body chemosensory discharge, that the adenosine receptor is an A_2-receptor (Monteiro & Ribeiro 1987).

Theophylline. The excitatory effect of CADO was compared before and in the presence of theophylline in a dose that did not cause modifications in respiration. Theophylline shifted the dose-response curve for CADO to the right, (Monteiro & Ribeiro 1987).

Manipulation of endogenous adenosine at the carotid body level.

The effect of adenosine deaminase inhibition. The adenosine deaminase inhibitor, erythro-9-(2-hydroxy-3-nonyl)adenine [EHNA (50-500 nmol.min^{-1} i.c. for 1 min)] increased both tidal volume (V_T) and frequency (f) of the respiratory movements. This effect was usually seen in the first 20 s after starting the EHNA infusions, maximal in the next 20 s, and disappeared in 2-3 min after finishing the EHNA infusions. This effect was associated with a small decrease in BP, and prevented by adenosine deaminase (10 I.U.min^{-1}). This enzyme alone did not modify V_T, f, HR or BP. EHNA in a dose (10 nmol) which did not cause detectable modifications on respiration, BP or HR, shifted to the left the dose-response curve for adenosine by a factor of 10, suggesting that EHNA potentiates the excitatory effect of exogenous adenosine on respiration probably by inhibition of adenosine deaminase. Theophylline and 1,3-dipropyl-8-(p-sulfophenyl)xanthine (DPP-8SPX) in doses that did not modify ventilation antagonized the excitatory effect of EHNA on respiration.

The effect of adenosine uptake inhibition. Intracarotid (i.c.) infusions of dipyridamole (19.8-198 nmol.min^{-1} during 1 min). caused increases in V_T and f. The excitatory effect of dipyridamole was observed in the first 20 s that followed the beginning of its i.c. infusion, was maximal in the last 10 s of the infusion and lasted, depending on the dose, for more than 5 min after the end of the infusions. After bilateral section of carotid sinus nerves the effect of dipyridamole virtually disappeared. Small decreases in BP were observed after infusion of high (148-198 nmol) doses of dipyridamole.

The effect of adenosine receptor antagonism. The alkylxanthines, theophylline and DPP-8SPX, a water soluble potent adenosine antagonist (Daly et al. 1985) decreased respiratory ventilation. After section of the carotid sinus nerve this inhibitory effect of DPP-8SPX disappeared, and with high (>1 umol) doses stimulated respiration. Whether this excitatory effect relates to central nervous system stimulation needs to be investigated.

3. The action of adenosine analogues on ventilation mediated by the central nervous system.

It has been described that i.v. L-PIA depresses respiration in vagotomized and glomectomized cats, i.e. animals in which the carotid sinus nerves have been cut (Eldridge et al. 1984), as well as when administered i.p. in vagotomized but not glomectomized rats (Wessberg et al. 1985). A depression caused by L-PIA (100 nmol) in rats was also observed 20 min after its i.v. injection, an effect which remained after section of the carotid sinus nerve (Monteiro & Ribeiro 1987). This suggests that the depression caused by L-PIA on respiration has a central origin. To test this further we administered CADO i.c.v. to cats, an animal species with chemoreceptors highly sensitive to CADO i.c., and we observed that CADO (166 nmol) decreased f, an effect which was maximal 20 min after the injection of the nucleoside. This effect of CADO was reversed by theophylline (5.5 umol i.c.v.).

CONCLUSIONS

The results reported in the previous sections suggest that adenosine may be a regulator of respiration at both peripheral and central levels. The peripheral regulation related to the excitatory action of adenosine on respiration is exerted mainly at the carotid body level, where adenosine excites carotid body chemosensory activity. This regulation seems to be physiologically important in adults. The central regulation is related to the depressant effects of adenosine on respiration and might be relevant in preterms, neonates, and glomectomized animals.

REFERENCES

Aubier, M., De Troyer, A., Sampson, M., Macklem, P.T. & Roussos, C., 1981, New England Journal of Medicine, 305, 249.

Biaggioni, I., Olafsson, B., Robertson, R., Hollister, A.S. & Robertson, D., 1986, Pflugers Archiv, 407, P109.

Buss, D.C., Routledge, P.A. & Watt, A.H., 1986, British Journal of Pharmacology, 88, 412 P.

Daly, J.W., Padgett, W., Shamim, M.T., Butts-Lamb, P. & Waters, J., 1985, Journal of Medicinal Chemistry, 28, 487.

Eldridge, F.L., Millhorn, D.E. & Kiley, J.P., 1984, Brain Research, 301, 273.

Eyzaguirre, C. & Fidone, S.J., 1980, American Journal of Physiology, 239, C135.

Hedner, T., Hedner, J., Jonason, J. & Wessberg, P., 1984,

European Journal of Respiratory Diseases, 65, 153.
Henon, B.K. & McAfee, D.A., 1983, Journal of Physiology, 336, 607.
Lagercrantz, H., Yamamoto, Y., Fredholm, B.B., Prabhakar, N.R. & von Euler, C., 1984, Pediatric Research, 18,387.
Mann, J.S., Holgate, S.T., Renwick, A.G. & Cushley, M.J., 1986, Journal of Applied Physiology, 61, 1667.
Maxwell, D.L., Fuller, R.W., Nolop, K.B., Dixon, C.M.S. & Hughes, J.M.B., 1986, Journal of Applied Physiology, 61, 1762.
McQueen, D.S., 1977, Journal of Physiology, 273, 515.
McQueen, D.S. & Ribeiro, J.A., 1981, British Journal of Pharmacology, 74, 129.
McQueen, D.S. & Ribeiro, J.A., 1983, British Journal of Pharmacology, 80, 347.
McQueen, D.S. & Ribeiro, J.A., 1986, British Journal of Pharmacology, 88, 615.
Millhorn, D.E., Eldridge, F.L., Kiley, J.P. & Waldrop, T.G., 1984, Respiration Physiology, 57, 331.
Monteiro, E.C. & Ribeiro, J.A., 1987, Naunyn-Schmiedeberg's Archives of Pharmacology, 335, 143.
Pauwels, R. & van der Straeten, M., 1986, Archives Internationales de Pharmacodynamie et Therapie, 280, 229.
Reid, P.G., Watt, A.H., Routledge, P.A. & Smith, A.P., 1987, British Journal of Clinical Pharmacology, 23, 331.
Ribeiro, J.A. & McQueen, D.S., 1983, On the neuromuscular depression and carotid chemoreceptors activation caused by adenosine. In Physiology and Pharmacology of adenosine derivatives, edited by J.W. Daly, Y. Kuroda, J.W. Phillis, H. Shimizu & M. Ui (Raven Press, N.Y.), p. 179.
Ribeiro, J.A. & Sebastião, A.M., 1986a, Pflugers Archiv,407, P20.
Ribeiro, J.A. & Sebastião, A.M., 1986b, Progress in Neurobiology, 26,179.
Runold, M., Lagercrantz, H. & Fredholm, B.B., 1986, Journal of Applied Physiology, 61, 255.
Viires, N., Aubier, M., Murciano, D., Marty, C. & Pariente, R., 1986, American Review of Respiratory Diseases, 133, 1060.
Watt, A.H. & Routledge, P.A., 1985, British Journal of Clinical Pharmacology, 20, 503.
Watt, A.H., Buss, D.C. & Routledge, P.A., 1987a, Life Sciences, 40, 29.
Watt, A.H., Reid, P.G., Stephens, M.R. & Routledge, P.A., 1987b, British Journal of Clinical Pharmacology, 23, 486.
Wessberg, P., Hedner, J., Hedner, T., Persson, B. & Jonason, J., 1985, European Journal of Pharmacology, 106, 59.

Adenosinergic Modulation of Brainstem Mechanisms Involved in Cardiovascular Control

Robin A. Barraco

Department of Physiology
Wayne State University Medical School
540 E. Canfield
Detroit, Michigan 48201 U.S.A.

INTRODUCTION

The physiological and pharmacological actions of adenosine receptor agonists and antagonists are well documented (Daly 1985; Dunwiddie 1985; Snyder 1985). It follows that if adenosine is an endogenous modulator of neural substrates mediating behaviors such as locomotion, antinocisponsiveness, arousal levels, sleep and ingestive behaviors (Barraco 1985), it is possible that adenosine may also influence CNS areas regulating vegetative behavior such as respiration and cardiovascular function. Indeed, a number of laboratories have shown that injections of adenosine analogs into the brain ventricles produce marked respiratory depression (Eldridge et al. 1984; Lagercrantz et al. 1984) which is often accompanied by hypotension and bradycardia (Hedner et al. 1982; Mueller et al. 1984; Wessberg et al. 1985). It has also been shown that unilateral injections of adenosine analogs into the lateral cerebral ventricle of rats produce dose-related reductions in blood pressure and heart rate and these centrally-mediated cardiovascular effects were antagonized by parenteral injections of caffeine (Barraco et al. 1984). In a more recent study, microinjections of 5'-N-ethylcarboxamidoadenosine (NECA) were given unilaterally into specific regions adjacent to and within the ventricular system, in an effort to identify the neural site(s) mediating the cardiovascular responses to intraventricular injections of adenosine analogs (Barraco et al. 1986). The data showed that brainstem sites, especially the Nucleus Tractus Solitarius (NTS) and the posterior portion of the fourth ventricle in the region of the area postrema, were the most sensitive sites for the hypotensive and bradycardic actions of NECA. The purpose of the present study was to extend these findings on the central cardiovascular actions of adenosine analogs in the rat. In the first study, analogs of adenosine were injected into the posterior region of the fourth ventricle of rats and their dose-response effects on blood pressure and heart rate assessed. In addition, the antagonism of these responses by parenteral injections of caffeine was examined. Portions of this study have been reported (Barraco et al. 1987b). In the second study, the dose-response effects of adenosine on cardiovascular parameters were examined following discrete administration into the caudal NTS via microinjection.

STUDY I: MICROINJECTIONS OF ADENOSINE ANALOGS INTO THE FOURTH VENTRICLE

Materials and Methods

Adult male Sprague-Dawley rats (300-350 g) were stereotaxically implanted with unilateral indwelling stainless-steel guide cannulae for injection into the posterior region of the fourth ventricle (P4V) (AP:12.8 mm (Bregma); ML:0.1 mm; DV:7.8 mm from horizontal skull surface) and allowed to recover following surgery. For acute blood pressure experiments, rats were anesthetized with urethane (1.0 g/kg i.p.). Blood pressure and heart rate recordings were made as described in detail previously (Barraco et al. 1986; 1987b). Cardiovascular responses were recorded for a minimum period of 60 min. after each injection. Cardiovascular parameters were analyzed by the Kruskal-Wallis ranked one-way analysis of variance and Mann-Whitney U post-hoc comparisons. NECA and (-)-N-(1-methyl-2-phenylethyl)adenosine (L-phenylisopropyladenosine, L-PIA) were injected in a volume of 0.4μl over a 30-s period. Saline (0.4μl) was used for control injections. In order to examine adenosine receptor antagonism, in separate experiments, caffeine (160 μmols/kg) was injected i.p. (0.1 ml) 30 min. prior to the injection of the highest doses of adenosine analogs. The placement of the cannula was verified microscopically following each experiment.

Results

The control value for mean arterial blood pressure (P_A) and heart rate (HR) prior to injections was 95.7 ± 3.8 mm Hg and 391 ± 12 beats/min., respectively. Control injections of saline (n = 10) had no observable effect on any of the cardiovascular parameters examined. The effects of microinjections of NECA into the posterior region of the fourth ventricle on P_A and HR are shown in Fig. 1. Overall analysis of variance indicated that NECA caused significant dose-related reductions in blood pressure (H = 26.88, df = 4, $P<0.001$) and heart rate (H = 24.54, df = 4, $P<0.001$). Significance levels, determined for individual doses by post-hoc comparisons with controls, are indicated by asterisks in Fig. 1. The effects of microinjections of L-PIA into the posterior region of the fourth ventricle on P_A and HR are shown in Fig. 2. Overall analysis of variance indicated that L-PIA caused significant dose-related reductions in P_A (H = 17.30, df = 4, $P<0.001$) and HR (H = 12.13, df = 4, $P<0.05$). Significance levels, determined for individual doses by post-hoc comparisons with controls, are indicated by asterisks in Fig. 2. Pretreatment with caffeine antagonized the maximal hypotensive and bradycardic effects of both analogs at the highest doses and at a time (30 min. after i.p. injections) when caffeine had no observable effects on any of the cardiovascular parameters examined.

Discussion

The results of this study show that microinjections of adenosine analogs into the posterior portion of the fourth ventricle in the region of the area postrema produced dose-related reductions in blood pressure and heart rate. Related findings were reported in a previous study examining the cardiovascular effects of intracerebroventricular (i.c.v.) injections of adenosine analogs in the rat wherein NECA and L-PIA also produced dose-related reductions in blood pressure and heart rate. However, in the present study the doses of analogs required to elicit significant hypotension were 100 - 1000 fold less than those required to depress blood pressure to a similar magnitude following i.c.v. injections (Barraco et al. 1986) or following parenteral administration of NECA and L-PIA by direct injection into the femoral vein (Barraco et al. 1987a). In addition, in agreement with previous findings using i.c.v. injections (Barraco et al. 1986), the results from the present study show that the central cardiovascular effects of both NECA and L-PIA were antagonized by parenteral injections of caffeine.

In this study, it is probable that the adenosine analogs acted at loci within or adjacent to the fourth ventricle in the immediate vicinity of the injection site (Barraco et al. 1987b). It has long been recognized that neurons within the lower brainstem are critical for the tonic and reflex regulation of systemic arterial pressure (Reis et al. 1984). Moreover, the major relay nucleus involved in baroreceptor reflex function, the NTS, especially the larger medial portion, forms a portion of the wall of the fourth ventricle just rostral to the area postrema, and it has already been

Fig. 1 (left). The cardiovascular effects of microinjections of NECA into the P4V and pretreatment with caffeine 30 min prior to injection of analog at the 0.50 nmol/kg dose. Values represent the mean ± S.E.M. percent reduction in control values: *differs from controls at $P < 0.01$; **pretreatment with caffeine in combination with analog, differs from analog alone at $P < 0.01$ ($n = 6$).

Fig. 2 (right). The cardiovascular effects of microinjections of L-PIA into the P4V and pretreatment with caffeine 30 min prior to injection of analog at the 3.5 nmol/kg dose. Values represent the mean ± S.E.M. percent reduction in control values: *differs from controls at $P < 0.01$; **pretreatment with caffeine in combination with analog, differs from analog alone at $P < 0.025$ ($n = 6$).

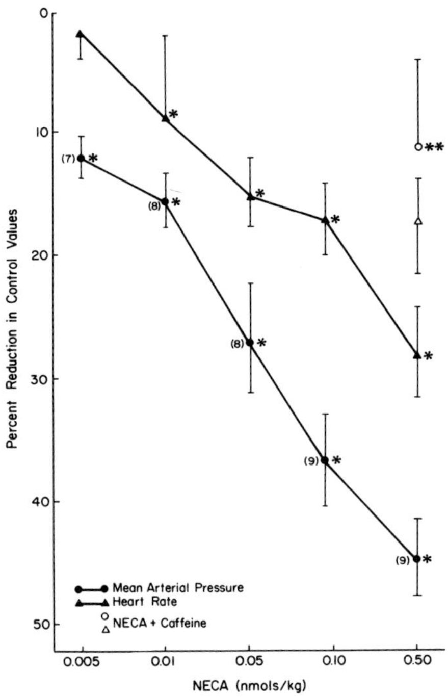

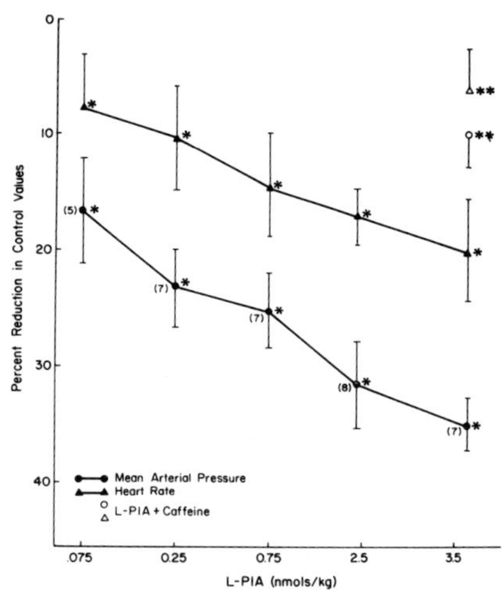

shown that microinjections of NECA into the NTS can elicit depressor responses (Barraco et al. 1986). Another prominent possibility as a site of action within the fourth ventricle is the area postrema (AP) which has been proposed to play a role in central cardiovascular control (Barnes et al. 1984). The AP is known to have, for example, rich afferent and efferent connectivity with established bulbopontine cardiovascular control areas, especially the NTS, the ventrolateral medulla, the dorsal motor nucleus of the vagus nerve and the nucleus ambiguus (Leslie 1986). Indeed, in the rat the ventral portion of the rostral AP forms a boundary with the neighboring area of the NTS (Takeuchi & Sano 1983).

STUDY II: MICROINJECTIONS OF ADENOSINE INTO THE NTS

Materials and Methods

Animals, anesthesia and recording of cardiovascular parameters were the same as in the first study. Thereafter, the subject was mounted in a cranial stereotaxic apparatus and a limited occipital craniotomy was conducted via electrocautery of neck muscles and then delicate dissection to reveal the caudal medulla in the region of the obex. Microinjections were made stereotaxically into brainstem sites with beveled glass micropipettes with a tip diameter of 60-80 μm. The entire microinjection procedure was conducted under a dissecting microscope. The coordinates for microinjections into the caudal NTS were: AP:-0.2 mm (from obex); ML: 0.3 mm; DV: 0.3 mm from dorsal surface of brainstem. Drug or control solutions (ACF) were injected in 100 nl volume over a 15 sec period. Cardiovascular parameters were recorded for a minimum of 30 min. Experimental values were calculated as the mean ± S.E.M. percent of control values determined prior to injection for each animal. Mean arterial blood pressure (P_A) and heart rate (HR) were measured prior to injection (control values), at the time of maximum effect (T_m) and at the time of recovery (T_r). Data were analyzed by Kruskall-Wallis ranked one-way analysis of variance and Mann-Whitney U post hoc comparisons. At the completion of each experiment, brains were processed histologically in 47 μ sections to verify injection sites.

Results and Discussion

The effects of microinjections of varying doses of adenosine on P_A and HR expressed as percent reduction in control values are shown in Fig. 3. Control injections of ACF (n = 7) had no observable effect on any of the cardiovascular parameters examined. Overall analysis of

Fig. 3 (left). The effects of adenosine on P_A and HR following microinjection into the caudal NTS. Values represent the mean ± S.E.M. percent reduction in control values which were established prior to injections. Numbers in parentheses indicate the number of injections for each drug group. Intersection at the ordinate = ACF controls (n = 7). Significance levels determined by post-hoc comparisons: *differs from vehicle controls at $P < 0.001$.
Fig. 4 (right). Actual tracings of blood pressure and heart rate following microinjections of adenosine into the caudal NTS at three separate doses: a) 0.05 nmoles/kg; b) 1.0 nmoles/kg; c) 10.0 nmoles/kg.

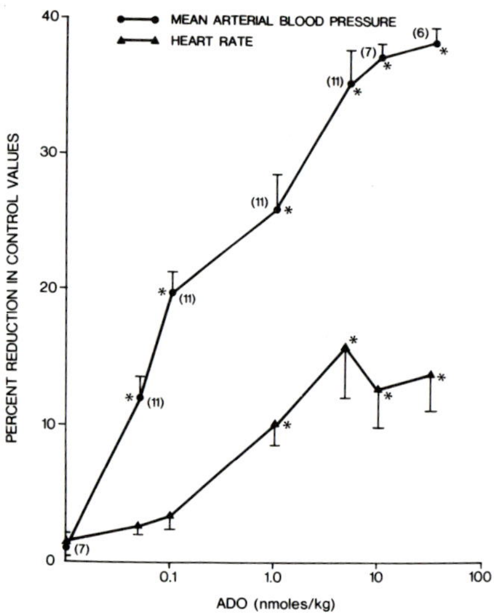

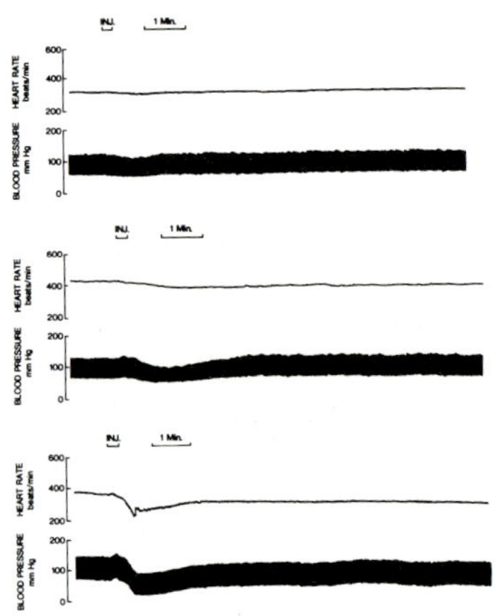

variance indicated that adenosine produced significant dose-related reductions in P_A ($H = 41.30$, $df = 5$, $p < 0.001$) and HR ($H = 29.25$, $df = 5$, $p < 0.001$). Significance levels, determined for individual doses by post hoc comparisons with ACF controls, are indicated by asterisks in Fig. 3. In Fig. 4 are shown the actual tracings of blood pressure and heart rate following microinjection of three separate doses of adenosine. The T_m for the hypotensive response was similar for all doses and occurred usually within 120 sec after the completion of the injection (mean T_m for all doses = 92.4 ± 9 sec). In contrast, the T_R of P_A to control values showed significant dose-dependent effects. For example, at the highest dose, the T_R = 7.14 ± 1.6 min. Overall analysis of variance indicated that adenosine produced significant dose-related increases in T_R for the hypotensive response ($H = 31.16$, $df = 5$, $p < 0.0005$).

The results of the second study show that microinjections of adenosine into the caudal NTS produced dose-related reductions in blood pressure and heart rate. This is especially noteworthy since the NTS is a major relay nucleus for afferent input from baroreceptors and thereby plays a critical role in central cardiovascular control (Brody et al. 1986). The NTS has been shown to have a complex morphology in addition to possessing a site-specific topographical distribution of a wide array of neurotransmitters and putative neuromodulators (Leslie 1985). Indeed, the discrete administration of many of these neuroactive substances into the NTS has been shown to elicit potent cardiovascular responses. Receptors for many of these substances have also been identified in the NTS (Moss et al. 1986) including receptors for adenosine (Lewis et al. 1981). In this regard, it is significant that the caudal NTS also exhibits the highest density of adenosine uptake sites of any region in the rat brain (Bisserbe et al. 1985).

In conclusion, the data from these experiments support the possibility that adenosine plays a neuromodulatory role in brainstem mechanisms involved in cardiovascular regulation.

This research was supported by grants from NSF (R1186-04084), NIH (RR-08167-09) and the American Heart Association of Michigan.

REFERENCES

Barraco, R.A., Aggarwal, A.K., Phillis, J.W., Moron, M.A. & Wu, P.H., 1984, Neuroscience Letters, 48, 139.

Barraco, R.A., 1985 behavioral action of adenosine analogues. In Purines: Pharmacology and Physiological Roles, edited by T. Stone (MacMillan Press), p. 27.

Barraco, R.A., Phillis, J.W., Campbell, W.R., Marcantonio, D.R. & Salah, R.S., 1986, Neuropharmacology, 25, 675.

Barraco, R.A., Marcantonio, D.R., Phillis, J.W. & Campbell, W.R. 1987a, General Pharmacology, 18, 405.

Barraco, R.A., Campbell, W.R., Schoener, E.P., Shehin, S.E. & Parizon, M., 1987b, Brain Research (in press).

Barnes, K.L., Ferrario, C.M., Chernicky, C.L. & Brosnihan, K.B., 1984, Federation Proceedings, 43, 2959.

Bisserbe, J.C., Patel, J. & Marangos, P.J., 1985, Journal of Neuroscience, 5, 544.

Brody, M.J., Alper, R.H., O'Neill, T.P. & Porter, J.P., 1986 central neural control of the cardiovascular system. In Pathophysiology of Hypertension-Regulatory Mechanisms, Vol. 8, edited by A. Zanchetti and R.C. Tarazi (Handbook of Hypertension) (Elsevier Science Publishers), p. 1.

Daly, J.W., 1985 adenosine receptors. In Advances in Cyclic Nucleotide and Protein Phosphorylation Research, edited by D. Cooper & K. Seamon (Raven Press), p. 29.

Dunwiddie, T.V., 1985, International Review of Neurobiology, 27, 63.

Eldridge, F.L., Millhorn, D.E. & Kiley, J.P., 1984, Brain Research, 59, 1428.

Hedner, T., Hedner, J., Wessberg, P. & Jonasson, J., 1982, Neuroscience Letters, 33, 147.

Langercrantz, H., Yamamoto, Y., Fredholm, B.B., Prabhakar, N.R. & Euler, C., 1984, Petiatric Research, 18, 387.

Leslie, R.A., 1985, Neurochemistry International, 7, 191.

Leslie, R.A., 1986, Cellular and Molecular Neurobiology, 6, 95.

Lewis, M.E., Patel, J., Edley, S. Moon & Marangos, P.J., 1981, European Journal of Pharmacology, 73, 109.

Moss, I.R., Denavit-Saubie, M., Eldridge, F.L., Gillis, R.A., Herkenham, M. & Lahiri, S., Federation Proceedings, 45, 2133.

Mueller, R.A., Winderlov, E. & Breese, G.R., 1984, Pharmacology Biochemistry and Behavior, 21, 289.

Reis, D.J., Granata, A.R., John, T.H., Ross, C.A., Ruggiero, D.A. & Park, D.H., 1984, Hypertension, 6 (Suppl. II), 379.

Snyder, S.H., 1985, Annual Review Neuroscience, 8, 103.

Takeuchi, Y. & Sano, Y., 1983, Anatomy and Embryology, 167, 311.

Wessberg, P., Hedner, J., Hedner, T., Persson, B. & Jonasson, J., 1985, European Journal of Pharmacology, 106, 59.

ADENOSINE AND CARDIOVASCULAR CONTROL

David Robertson, Italo Biaggioni, Ching-Jiunn Tseng

Departments of Medicine and Pharmacology
Vanderbilt University
Nashville, TN 37232 USA

In 1978 we began a series of studies (Robertson et al., 1978, 1981, 1984) of the clinical pharmacology of caffeine in normal subjects acutely and chronically and also in hypertensive subjects. Modest acute effects of caffeine to raise blood pressure were observed, but chronic effects of caffeine were too small to be detected.

We subsequently used disease models to determine the role of the sympathetic nervous system in mediating the effects of caffeine and concluded that an intact sympathetic nervous system was not required for caffeine's cardiovascular effect (Onrot et al., 1985). These studies, however, did not rule out a contributory role for the sympathetic nervous system and we continue to pursue this potential mechanism of caffeine's effect. In ancillary studies, we demonstrated that caffeine was a very effective treatment of postprandial hypotension and its associated dizziness in patients with autonomic impairment (Onrot et al., 1985, 1986; Robertson et al., 1981). Caffeine has subsequently become a routine therapy of such patients.

With increasing evidence that adenosine was involved in the mechanism of action of caffeine, we turned to adenosine mechanisms hoping they would provide information about caffeine's actions (Biaggioni et al., 1985, 1986a, 1986b, 1987a, 1987b). Adenosine had been reported to be an effective hypotensive agent in anesthetized patients and had been reported to be effective in treating supraventricular tachycardia. It had also been proposed as a mediator of blood flow autoregulation. Its effect on cardiovascular control was thus of major importance, and we studied its clinical pharmacology, particularly in terms of hemodynamics and sympathetic nervous system effects.

The cardiovascular and respiratory effects of intravenous adenosine were studied in conscious normal volunteers. Bolus injections of adenosine increased systolic and diastolic pressures initially (+15 and +13 mmHg after 100 µg/kg) followed by a subsequent reduction in systolic and diastolic pressures (-12 and -16 mmHg). Heart rate increased during trough blood pressure (R-R interval shortening of 298 msec after 100 µg/kg). Adenosine steady-state infusions increased heart rate (+30 bpm during 140 µg/kg/min), systolic pressure (+16 mmHg), and pulse pressure (+21 mmHg), but decreased diastolic pressure slightly (-5 mmHg), resulting in no significant change in mean arterial pressure. Adenosine stimulated respiration, resulting in decreased pCO_2 (41 to 31 mmHg), increased pO_2 (101 to 113 mmHg) and increased pH (7.42 to 7.50). The increased ventilation was not explained by bronchoconstriction, hypotension or hypoxia. The observed pressor and tachycardic effects are mediated through reflex autonomic mechanisms, since they are completely abolished in patients with severe autonomic failure. The hemodynamic and respiratory effects of adenosine appear to be due in part to chemoreceptor stimulation since we found that adenosine was pressor when infused in the aortic arch proximal to the origin of the carotid arteries, but depressor when infused in the descending aorta. Such findings raise the possibility that adenosine is an endogenous modulator of respiration and cardiovascular control in man.

Our studies of caffeine in man (Robertson et al., 1978) and the more recent studies from Dr. Edwin K. Jackson's laboratory at our institution (Deray et al., 1986; Ohnishi et al, 1986) provide the strong suggestion that a link exists between renin and the effects of adenosine and caffeine. These latter experiments indicated that caffeine specifically exacerbated experimental renovascular hypertension in rats but did not worsen spontaneous (SHR) hypertension. Given the well-known effects of caffeine to alter alertness, we considered that adenosine might have important effects in the central nervous system and that these effects might help explain some of the observations made in our caffeine studies.

Autonomic cardiovascular control is achieved in the medulla oblongata through integration of input from higher centers with afferent input from receptors in chest and neck structures (Reis et al, 1984). This afferent input travels in the vagus and glossopharyngeal nerves and terminates in the nucleus tractus solitarii (NTS). NTS neurons have projections to many rostral and ventral sites, like the C1 area in the ventral lateral medulla (Ross et al, 1985). Stimulation of the NTS produces inhibition of sympathetic tone while stimulation of the C1 area leads to an increase in sympathetic activation (Granata et al, 1985).

Much progress has been made in the identification of neurotransmitters and neuromodulators in these medullary sites. Evidence has been brought forward in support of a role for L-glutamate (Dampney et al., 1982), GABA (Yamada et al., 1982), epinephrine (Granata et al, 1985) and neuropeptide Y (Everitt et al., 1984). It seemed likely to us that additional ones might also contribute to medullary cardiovascular control, and we considered that adenosine might be one of these. Specific receptors for adenosine have been found in the central nervous system (Daly et al., 1981) and appear to modulate intracellular levels of adenosine 3',5'-cyclic monophosphate (Ebersolt et al, 1983; Yeung et al., 1983). Tissue levels of adenosine in the central nervous system have been shown to increase during hypoxia (Winn et al., 1981) and seizures (Winn et al., 1980).

Figure 1: Effects of increasing doses of adenosine on blood pressure and heart rate. Site of injection: Nucleus Tractus Solitarii. • , change in heart rate, ▲ , ■ , change in diastolic and systolic blood pressure (Asterisk * denotes $p < 0.001$ by analysis of variance).

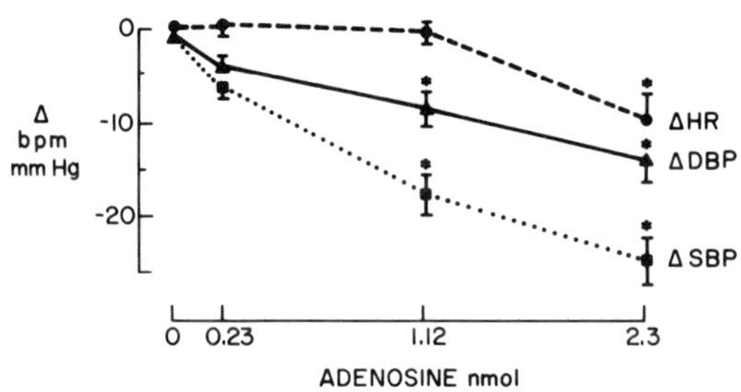

Adenosine inhibits the release of many neurotransmitters including acetylcholine (Ginsborg & Hirst, 1972; Vizi & Knoll, 1976; Murray et al., 1982), noradrenaline (Harms et al., 1978; Fredholm et al, 1985; Tung et al., 1987), dopamine (Michaelis et al, 1978), GABA (Harms et al., 1979) and L-glutamate (Dolphin & Archer, 1983), both in the central nervous system and in peripheral tissues. Adenosine also has a marked depressive action on the firing of neurons in virtually all regions of the central nervous system tested (Phillis et al, 1974). It has also been proposed that adenosine, or ATP, may act as a neurotransmitter or cotransmitter in some nerves (Fredhold & Hedqvist, 1980).

We studied the responses evoked by stereotaxically administered adenosine in critical brainstem sites of rats (Tseng et al., 1987). Microinjection (60 nl) of adenosine (0, 0.23, 1.12, 2.3 nmol in artificial cerebral spinal fluid) into the NTS produced a dose-related decrease in heart rate and in systolic and diastolic blood pressure (Fig. 1). Maximal changes occurred 90 seconds following injection. The injection of 2.3 nmol adenosine reduced heart rate by 10 ± 2 bpm and reduced blood pressure by 24/14 ± 2/2 mmHg (n=7, p<0.001). A somewhat greater effect was observed with injection into the area postrema. Heart rate in this site fell 18 ± 1 bpm while blood pressure fell 29/25 ± 4/3 mmHg (n=6, p<0.001). Similarly, microinjection of ATP (0.01 - 1.1 nmol) and its analogue β,γ-methylene ATP (0.01 - 1.1 nmol) were made into the NTS area. The injection of 1.1 nmol ATP reduced heart rate by 55 ± 2 bpm and reduced blood pressure by 34 ± 1 mmHg (n=6, p<0.001) . Injection of 1.1 nmol β,γ-methylene ATP decreased heart rate by 73 ± 2 bpm and reduced blood pressure by 40/38 ± 2/1 mmHg (n=6, p<0.001) (Fig. 2).

Figure 2: Effects of increasing doses of ATP on blood pressure and heart rate. Site of injection; Nucleus Tractus Solitarii. • , change in heart rate; ▲ ,■, change in diastolic and systolic blood pressure (Asterisk * denotes p<0.001 by analysis of variance).

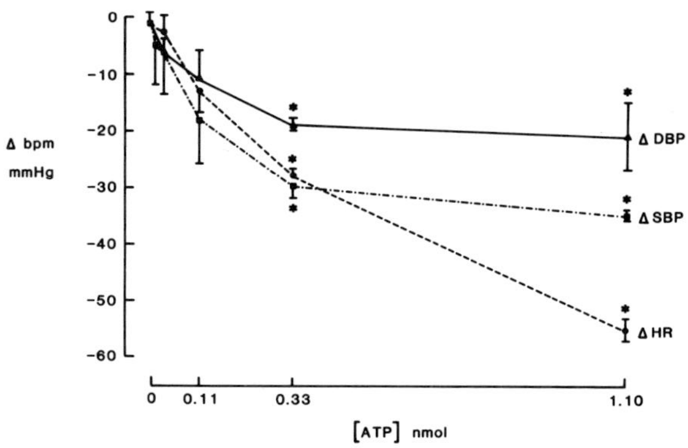

To determine the specificity of the effects of adenosine and ATP, we used the potent adenosine receptor antagonist, 1,3-dipropyl-8-p-sulfophenylxanthine (0.92 nmol) (Daly et al., 1985). The injection of this antagonist itself into the NTS had no effect on heart rate or blood pressure. However, it completely abolished the effect of further injections of adenosine or ATP in the NTS (Fig. 3), while leaving responsiveness to L-glutamate unimpaired (Fig.

4,5). Parallel studies showed similar results in the area postrema. We also found that adenosine deaminase (6 ng) locally injected into the NTS and AP completely blocked the effect of adenosine, ATP, and β,γ-methylene ATP. This further supports our view that ATP and β,γ-methylene ATP (or a closely related compound) were converted to adenosine and act through P_1 receptors.

Figure 3: Comparative effects of adenosine injected into the nucleus tractus solitarii (NTS), C1, and area postrema (AP), before and after 1,3-dipropyl-8-p-sulfophenylxanthine (DPSPX) an adenosine receptor antagonist. (Asterisk * denotes p < 0.001 by unpaired two-tailed T test.)

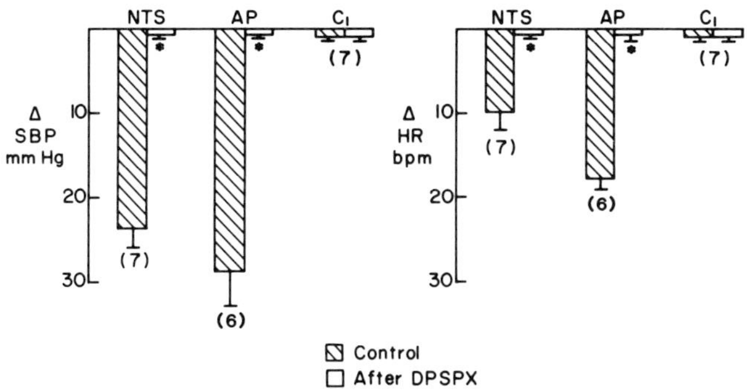

In contrast to the significant bradycardic and vasodepressor effects of adenosine and ATP in the NTS and area postrema, the injection of similar amounts of adenosine and ATP into the C1 area of the ventral lateral medulla was devoid of pharmacologic effect on blood pressure or heart rate.

Similar results were obtained in normotensives (WKY) and genetically hypertensive (SHR) rats. However, there was a significant difference between SHR and WKY in the magnitude of blood pressure depression in response to adenosine. This can be shown most clearly in the following data obtained following injection of 2.3 nmol of adenosine:

		n	SBP	DBP	HR
NTS	WKY	6	-23±2	-11±2	-12±1
	SHR	8	-8±1 *	-4±0 *	-10±2
AP	WKY	6	-27±3	-11±1	-26±7
	SHR	8	-9±1 *	-5±1 *	-11±0*

*p<0.05

Thus, adenosine is a potent depressor agent in both the area postrema and the nucleus of the solitary tract of WKY rats, but is much less potent in the spontaneously hypertensive rats. Such findings may indicate altered purinergic or purinoceptor function in SHR.

Figure 4: Representative tracing of the cardiovascular effects of microinjections of adenosine (2.3 nmol) into the area postrema, before and after 1,3-dipropyl-8-p-sulfophenylxanthine (DPSPX) (0.92 nmol). The upper register displays blood pressure and the lower register displays beat-to-beat heart rate.

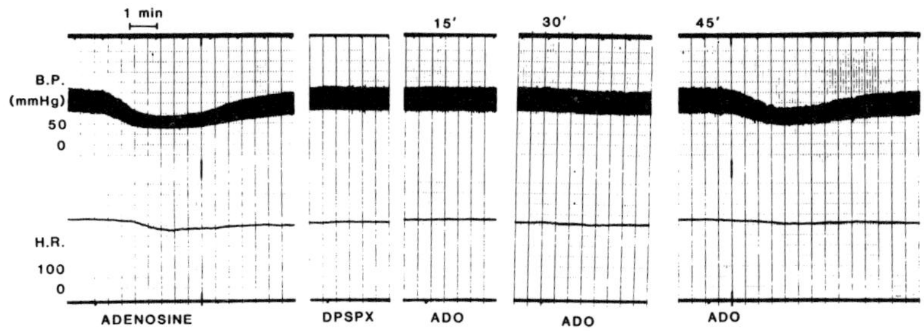

Since the identification of the NTS as an important nucleus in cardiovascular control, a variety of putative neurotransmitters and drugs have been tested for their potency in altering blood pressure and heart rate in this site. It is now known that noradrenaline, acetylcholine, adrenaline, L-glutamate, dopamine and other agents reduce blood pressure and heart rate when injected into the NTS. In contrast, certain agents such as angiotensin II (Casto & Phillips, 1984), vasopressin (Matsuguchi et al., 1984), and yohimbine (Tung et al., 1983) have been shown to raise blood pressure and heart rate when injected into this site.

The presence of adenosine receptors in many sites in the central nervous system encouraged us to consider the possibility that this agent might also exert important effects on cardiovascular control. We and others have recently shown that adenosine increases heart rate and systolic blood pressure and that it stimulates respiration in man (Biaggioni et al., 1986b). We believe these actions are due to adenosine-induced carotid body chemoreceptor activation (Biaggioni et al, 1987b). In contrast to these peripheral actions of adenosine, intracerebroventricular injection of adenosine has a depressor effect on respiratory control (Eldridge et al., 1985). Furthermore, the injection of the adenosine receptor agonists, 5'-N-ethylcarboxamido adenosine (NECA) and (-) -N^6-CR-phenylisopropyl adenosine (PIA) has recently been shown to exert substantial depressor effects in some medullary sites, including the NTS and the area postrema (Barraco et al., 1986).

Figure 5: Representative tracing of the cardiovascular effects of microinjection of β,γ-methylene ATP (1.1 nmol) into the Nucleus Tractus Solitarii, before and after 1,3-dipropyl-8-p-sulfophenylxanthine (DPSPX) (0.92 nmol). The upper register displays blood pressure and the lower register displays beat-to-beat heart rate.

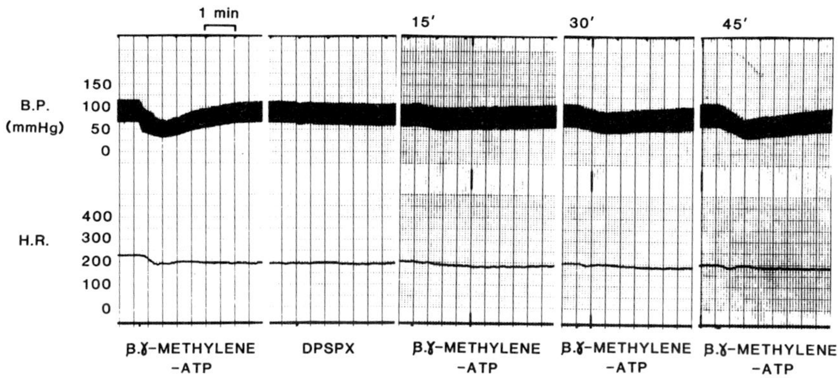

The fact that adenosine has such potent hypotensive actions in the area postrema (Tseng et al., 1987b, 1987c, 1987d) is especially interesting, since this area is essential in the development of hypertension in two kidney one clip rats (Fink et al., 1987) and caffeine, an adenosine receptor antagonist, has been shown to enhance the pressor effect of angiotensin II in this model of hypertension (Ohnishi et al., 1986, 1987).

Adenosine and ATP both reduce blood pressure and heart rate when injected into the NTS and the area postrema. The fact that an adenosine receptor antagonist blocked the depressor effects of adenosine but not that of L-glutamate is supportive evidence for the presence of specific adenosine receptors capable of modulating blood pressure in these medullary centers. Moreover, ATP was considerably more potent on a molar basis in eliciting these responses than adenosine itself, suggesting that an ATP receptor (a P_2 purinergic receptor) mediates these effects. However, since in our studies the effects of ATP were effectively blocked by the adenosine receptor antagonist, it is possible that adenosine receptors are mediating the responses observed. Adenosine is known to be very rapidly taken up into cells. When it is thus removed from the membrane receptor sites at which many of its effects are mediated, it cannot, of course, continue to exert its pharmacologic effect. It is therefore possible that the injection of adenosine into the nucleus of the solitary tract is followed by a relatively rapid transit of adenosine to intracellular sites. ATP, which may not be so readily transported into the cells, may thus penetrate more effectively to sites distant from the injection needle and release adenosine locally onto membrane receptors in those sites. This latter mechanism could account for the greater potency of ATP than adenosine, simply on the basis of its improved delivery of adenosine to its relevant receptor sites.

It is noteworthy that the administration of an adenosine antagonist into the NTS or AP did not itself exert a major effect on baseline blood pressure in our model. Since anesthesia greatly alters prevailing autonomic tone, we believe that different results might be obtained in the awake state or with use of alternative anesthetic agents. The mechanism of the differential effect in the spontaneously hypertensive rat and the question of whether renovascular hypertensive animals would be similar are issues that we propose to investigate further.

CONCLUSION

Adenosine has diverse effects on the cardiovascular system. In addition to its direct effect to dilate most arterioles, it acts on chemoreceptor structures in the distribution of the carotid artery to activate the sympathetic nervous system and enhance respiration. With adenosine administration intravenously into awake human subjects, this latter effect is dominant, and a hypotensive effect can be demonstrated only at the highest infusion rates.

Studies in which adenosine was injected into various medullary sites indicate that the purine elicits substantial hypotension and bradycardia in the *area postrema* and the *nucleus tractus solitarii* in both Sprague-Dawley and Wistar-Kyoto rats (WKY). This effect is markedly attenuated in spontaneously hypertensive rats. The reason for this striking difference in central adenosine sensitivity in the SHR and WKY rats is not yet clear.

ACKNOWLEDGEMENTS

Supported in part by HL34021 and the International Life Sciences Institute. Dr. Robertson is a Burroughs Wellcome Scholar in Clinical Pharmacology. Dr. Biaggioni is a Clinical Associate Physician on the Elliott V. Newman Clinical Research Center.

REFERENCES

Barraco RA, Phillis JW, Campbell WR, Marcantonio PR, Salah RS. 1986, Neuropharmacology 25:675.

Biaggioni IO, Jackson EK, Jackson K, Onrot J, Hollister AS, Robertson D. 1985, Clinical Research, 33:279A.

Biaggioni I, Olafsson B, Robertson RM, Hollister AS, Robertson D. 1986a, Pflugers Archives European Journal of Physiology, 407:S54.

Biaggioni I, Onrot J, Hollister AS, Robertson D. 1986b, Life Sciences, 39:2229.

Biaggioni I, Bennett B, Patel B, Jackson EK, Robertson D. 1987a, Clinical Research, 35:373A.

Biaggioni I, Olafsson B, Robertson RM, Hollister AS, Robertson D. 1987b, Circulation Research, in press.

Casto R, Phillips MI. 1984, American Journal of Physiology 246:R811.

Daly JW, Bruns RF, Snyder SH. 1981, Life Sciences, 28:2083.

Daly JW, Padgett W, Shamim MT, Butts-Lamb P, Waters J. 1985, Journal of Medicinal Chemistry 28:487.

Dampney RA, Goodchild AK, Robertson LG, Montgomery W. 1982, Brain Research, 249:223.

Deray G, Branch RA, Ohnishi A, Herzer WA, Jackson EK. 1986, European Journal of Pharmacology, in press.

Dolphin AC, Archer ER. 1983, Neuroscience Letters, 43:49.

Ebersolt C, Premont J, Prochiantz A, Perez M, Bockaert J. 1983, Brain Research, 267:123.

Eldridge FL, Millhorn DE, Kiley JP. 1985, Journal of Applied Physiology, 59:1428.

Everitt BJ, Hökfelt T, Terenius L, Tatemoto K, Mutt V, Goldstein M. 1984, Neuroscience, 11:443.

Fink GD, Bruner CA, Mangiapane ML. 1987, Hypertension 9:355.

Fredholm BB, Hedqvist P. 1980, Biochemical Pharmacology, 29:1635.

Fredholm BB, Jonzon B, Lindgren E. 1983, Acta Physiology Scandinavia 515:7.

Ginsborg BL, Hirst GDS. 1972, Journal of Physiology, 224:629.

Granata AR, Ruggiero DA, Park DH, Joh TH, Reis DJ. 1985, American Journal of Physiology, 248:H547.

Harms HH, Wardeh T, Mulder AH. 1978, European Journal of Pharmacology, 49:305.

Harms HH, Wardeh G, Muldor AH. 1979, Neuropharmacology, 18:577.

Matsuguchi H, Sharabi FM, Gordon FJ, Johnson AK, Schmid PG. 1984, Neuropharmacology 21:687.

Michaelis ML, Michaelis EK, Myers SL. 1979, Life Sciences

24:2083.

Murray TF, Blaker WD, Cheney DL, Costa E. 1982, Journal of Pharmacology and Experimental Therapeutics, 222:550.

Phillis JW, Kostopoulos GK, Limacher JJ. 1974; Canadian Journal of Pharmacology, 52:1226.

Ohnishi A, Branch RA, Jackson EK, Biaggioni I, Deray G, Jackson EK. 1986, Journal of Clinical Investigation, 78:1045.

Ohnishi A, Li P, Branch RA, Holycross B, Jackson EK. 1987, Journal of Clinical Investigation, 80:13.

Onrot J, Goldberg MR, Biaggioni I, Hollister AS, Kincaid D, Robertson D. 1985, New England Journal of Medicine, 313:549.

Onrot J, Shaheen O, Biaggioni I, Goldberg MR, Feely J, Wilkinson GR, Hollister AS, Robertson D. 1986, Clinical Pharmacology and Therapeutics, 40:506.

Reis DJ, Granata AR, Joh TH, Ross CA, Ruggiero DA, Park DH. 1984, Hypertension, II-6:7.

Robertson D, Frolich JC, Carr RK, Watson JT, Hollifield JW, Shand DG, Oates JA. 1978, New England Journal of Medicine, 298:181.

Robertson D, Hollister AS, Kincaid D, Workman R Goldberg MR, Tung C-S, Smith B. 1984, American Journal of Medicine, 77:54.

Robertson D, Wade D, Robertson RM. 1981a, American Journal of Cardiology, 40:1048.

Robertson D, Wade D, Workman R, Woosley RL, Oates JA. 1981b, Journal of Clinical Investigation, 67:1111.

Ross CA, Ruggiero DA, Reis DJ. 1985; Journal of Comparative Neurology, 242:511.

Tseng, CJ, Appalsamy, M, Biaggioni, I, Robertson, D. 1987a, Clinical Research, 35:18A.

Tseng, CJ, Appalsamy, M, Biaggioni, I, Robertson, D. 1987b, Clinical Research, 35:382A.

Tseng, CJ, Appalsamy, M, Robertson, D. 1987c, Submitted.

Tseng CJ, Appalsamy M, Biaggioni I, Robertson D. 1987d, Hypertension, in press.

Tung CS, Chu K-M, Tseng C-J, Yin TH. 1987, Life Sciences, in press.

Tung CS, Onuora CO, Robertson D, Goldberg MR. 1983, Brain Research, 1983; 277:193.

Vizi ES, Knoll J. 1976, Neuroscience, 1:391.

Winn HR, Welsh JE, Rubio R, Berne RM. 1980; Circulation Research, 47:568.

Winn HR, Rubio R, Berne RM. 1981, American Journal of Physiology, 241:H235.

Yamada KA, Norman WP, Hamosh P, Gillis RA. 1982, Brain Research, 248:71.

Yeung S-MH, Green RD. 1984, Naunyn-Schmiedeberg's Archives of Pharmacology, 325:218.

IN VIVO EFFECTS OF INHIBITORS OF ADENOSINE UPTAKE

Herman Van Belle

Department of Biochemistry, Janssen Pharmaceutica,
B-2340 Beerse, Belgium

INTRODUCTION

At the end of a symposium which is devoted to the physiology and pharmacology of adenosine any introduction can hardly be original: the discovery of adenosine, the major steps in unravelling its many pharmacological effects and the state of the art in its receptorology must all have been and, indeed, have been introduced to the audience by the previous speakers. As to the reader, references to the major topics in adenosine research may be found in this volume, as well as in some recent reports of major meetings on the subject (Berne et al. 1983-R; Paton 1985-R; Stefanovich et al. 1985-R; Gerlach & Becker, 1987-R). One aspect of the story has been, and is still being treated in a rather stepmotherly way: in this symposium, for instance, the topic of nucleoside transport inhibition has been assigned the very last place and it is the only one out of 25 dealing with adenosine. Nearly the same occurred during earlier meetings. There may be several reasons for that, as will be discussed further on - it is not due to any ill will of organizers or editors. With this lecture I hope to provide sufficient material to stimulate more research on the potential usefulness of transport inhibitors as drugs, as well as to encourage the search for new and more appropriate molecules.

AGONISTS OR NUCLEOSIDE TRANSPORT INHIBITION

Nowadays several laboratories provide us with a variety of nucleoside analogues showing a whole range of potencies as adenosine agonists, as well as a range of selectivities towards the various adenosine receptors. Their application has resulted in a wealth of information on the physiology of adenosine. Their slow catabolism, when compared with adenosine, has aided considerably in the indexing of many important roles of adenosine. Because of a different physicochemical behaviour, they can reach regions - and elicit responses - where systemic adenosine has no access. Their specificity has led to the discovery and discrimination of the adenosine receptors, which makes them important adjuncts in the molecular characterization of these receptors.

Being splendid pharmacological tools does not necessarily imply that these agonists are also very useful drugs. The major problem may be the almost universal distribution of adenosine receptors all over the body. Even if agonists are able to discriminate completely between A_1 and A_2 receptors - not to mention receptors X, Y or Z, which remain to be discovered - their systemic application will result in several responses in different tissues. Complete tissue specificity has - at least to my knowledge - not yet been achieved. It also follows that prophylactic use can hardly be imagined: too many processes may become activated and, being nucleoside analogues, cytotoxic effects could be expected after long-term treatment.

Considering the role of adenosine as a "retaliatory metabolite" (Newby 1984; Fredholm et al. 1983) the use of specific nucleoside transport inhibitors is much more appealing and more "natural" than the application of adenosine agonists. First, adenosine formation does not seem to be a continuously on-going process (except, probably, its release from the endothelium to prevent platelet adhesion - Gerlach et al. 1987). It is a reaction of tissue to the imbalance between energy demand and oxygen supply. It follows that its "release" is <u>local</u> and <u>temporary</u> - once oxygen supply is sufficient, the production of adenosine stops immediately (the same holds true if adenosine is released from nerves, as discussed by previous speakers). Second, the "spreading" of adenosine from the ischaemic area to other tissues is largely prevented in the first place by the endothelial cells, which avidly take up and catabolise adenosine, and in "second-line" by uptake into the blood cells. Apparently, nature has taken precautions to inhibit the transmission of the signal, which adenosine is, to other, healthy parts in the body. That is exactly the opposite of what one gets by using adenosine agonists.

Because, as recently evidenced (Dendorfer et al. 1987; Van Belle et al. 1987), AMP is the first product released from ischaemic cells, nucleoside transport inhibitors will not affect this process. Neither do they enhance the release nor the subsequent dephosphorylation to adenosine in the interstitium. What they effectively block is the uptake of adenosine in the endothelial cells. The final result will be a prolonged presence of adenosine in the area "at risk". Assuming that adenosine constitutes some "defense" system against the (many) deleterious processes that threaten ischaemic tissue, nucleoside transport inhibition may accentuate this important function. It also follows that, provided the inhibitors are highly specific, these drugs by themselves should be devoid of any activity. They will act when - <u>and then only</u> - and where - <u>and there only</u> - adenosine is produced. Of course, excessive doses may "spare" every molecule of adenosine produced, thus raising the levels in circulation sufficiently high to trigger every adenosine-sensitive system.

THE EXISTING NUCLEOSIDE TRANSPORT INHIBITORS

Thus far, specific nucleoside transport inhibition is confined to only a very few families of structurally unrelated chemicals. Of the more than 6,500 different molecules we have thus far screened <u>in vitro</u>, not one showed any effect at a micromolar concentration. Nitrobenzylthioinosine (NBMPR or NBTI), a nucleoside analogue, is very potent but, in sharp contrast with its frequent use as a ligand for the transporter, the compound has barely been applied <u>in vivo</u>. Almost equally potent - at least <u>in vitro</u> - is Dilazep (Cormelian), which is used as a therapeutic agent in man. By far the most widely known is Dipyridamole (Persantin), which, as a nucleoside transport inhibitor is almost 10 times less potent - <u>in vitro</u> - than dilazep. The same applies to Mioflazine, a more recent analogue of lidoflazine, but, as a transport inhibitor, more potent (= dipyridamole) and more selective. Mioflazine has not yet been applied extensively in man, but experiments in dogs reveal a remarkable activity (for details see Van Belle 1985).

Because of some drawbacks, inherent in the existing drugs, it is hard to get a clear view on their possible therapeutic benefits from <u>in vivo</u> studies in animals or man. Before discussing several problems <u>in vivo</u>, one should stress the insensitivity of the nucleoside transporter <u>in rats</u> to the existing drugs. Because the rat is still the species of choice for many pharmacologists and because adenosine is acting in the rat as it is in all species, experiments set up by many with the existing drugs turned out to be very disappointing. On the other hand, the dog is very sensitive to nucleoside transport inhibitors, but as we found very recently, when given orally, neither dilazep nor dipyridamole were absorbed to any extent (as measured from the *ex vivo* effect on nucleoside transport). Only mioflazine and several of its analogues were orally active and very long-acting (Van Belle, in press). Some pharmacokinetic data in man also reveal a rather poor bioavailability and/or short half-lives for dilazep (Schaumlöffel & Prignitz 1972) and dipyridamole (Dresse et al. 1982). The very high doses and the multiple daily dosages (dipyridamole: 4 x 75 mg; dilazep 3 x 50 mg) point in the same direction (especially in view of the marked *in vitro* potency of these compounds - $\leqq 10^{-8}$M !). Dipyridamole is also a rather potent inhibitor of some phosphodiesterases, of particulate adenosine kinase and of adenosine deaminase and has a definite effect on the transport of glucose and choline (for references see Kenakin & Leighton 1985). It also inhibits phosphate fluxes through membranes (IC_{50} = 1.5 x 10^{-6} M in human erythrocytes; our unpublished results). Some of these activities may show up at the micromolar concentrations which have to be reached *in vivo* in man to affect nucleoside transport (see Dresse et al. 1982). In this respect, dilazep looks much more specific, although it shows definite Ca^{2+}-antagonistic properties (Nakagawa et al. 1986) and is the strongest antiarrhythmic of a whole series of drugs in the isolated *rat* heart (Sono et al. 1986). Thus far, mioflazine seems to be largely devoid of any "foreign" activity. It does, however, suffer from the same disadvantage as dipyridamole - not shared by dilazep - namely a strong binding to the plasma proteins in man (not in dogs !). This binding necessitates concentrations in plasma which are from 50 to 100 times the concentrations needed in *in vitro* plasma-free systems to exhibit nucleoside transport inhibition (Van Belle, in press).

NUCLEOSIDE TRANSPORT INHIBITION *IN VIVO*

As said before, effective nucleoside transport inhibition should accentuate all known pharmacological actions of adenosine - provided that under the conditions of the test adenosine is formed. These actions include, among others: CNS-effects such as sleep, sedation, antianxiety and anticonvulsant (Dragunow 1986) effects; cardiovascular effects, such as platelet antiaggregation, coronary vasodilation, ß-blockade and antiarrhythmia; inhibitory effects on lipolysis and on neutrophil activation and production of free radicals (for many references: see books mentioned in the Introduction).

In view of the many possible actions it is very surprising that in animal studies, as well as in man, the therapeutic use of nucleoside transport inhibitors is restricted almost exclusively to the cardiovascular system. Cardioprotection after ischaemia has been established in dog models after intravenous dilazep (a.o. Seth et al. 1986) or dipyridamole (a.o. Blumenthal et al. 1981) and after oral treatment with mioflazine (see Van Belle, 1985). If accumulation of adenosine is the common denominator - and tissue analysis points into this direction (Van Belle et al. 1986) - the protection could be explained by its antiaggregating effect (preventing platelet activation and no-reflow), its modulating effect of positive inotropism and chronotropism from ß-agonists (preventing complete exhaustion of energy stores), its effect on ß-agonist stimulated lipolysis (preventing excessive accumulation of free fatty acids, which may compromise membrane integrity), its effect on neutrophil activation (preventing the production of deleterious radicals) and its antiarrhythmic effects. In short, nucleoside transport inhibition may tackle the many problems arising in the ischaemic heart and for which, otherwise, a battery of other drugs would be required.

Cardioprotection in man, which is much more difficult to prove, is less well-documented and is still a matter of scepticism. Several studies, however, evidence a beneficial effect of dilazep in the treatment of ischaemic heart disease (see Marzilli et al. 1984) and of the combination of dipyridamole and aspirin (Klimt et al. 1986). The underlying mechanism is supposed to be the effect on platelet behaviour (see Harker & Fuster 1986) which, at first, and for dipyridamole, was ascribed to its inhibition of platelet phosphodiesterase. More recent studies in whole blood, including dilazep (which has no effect on phosphodiesterase), revealed that in fact nucleoside transport inhibition is the main factor involved (see Dawicki et al. 1985). More selective, orally active, and longer-acting nucleoside transport inhibitors are needed to prove the potential of this treatment for cardioprotection. The same applies for the elucidation of the so-called "steal" effect, which may overshadow the therapeutic use. Being well-documented for dipyridamole at huge intravenous doses (around 0.5 mg/kg i.v. $\simeq 2 \times 10^{-5}$ M in plasma !) and in patients with significant coronary artery disease (see Picano et al. 1987), factors other than nucleoside transport inhibition may well be responsible. If adenosine was the major culprit, one should expect the phenomenon to occur much more frequently with dilazep, which, in man, and given intravenously, is 50 to 100 times more potent an inhibitor of nucleoside transport (intrinsic potency combined with lack of protein binding). Marzilli et al. (1984) reported that in 18 patients no ECG-evidence of ischaemia could be found after 0.2 mg/kg/i.v., which is nearly half the dose of dipyridamole used to elicit the ST-segment changes in some patients with coronary artery disease.

Data on CNS-effects of nucleoside transport inhibitors are very scarce. Intracerebroventricular administration of dilazep and dipyridamole depressed locomotor activity in rats (!), but subsequent caffeine antagonized the effect of dilazep only (Phillis et al. 1986). Sollevi et al. (1987) mention the lack of anxiolytic or sedative actions of dipyridamole. Therefore it is very interesting that oral administration of mioflazine, even at fairly low doses, improved sleep in dogs (Wauquier et al. 1987). The antagonism by caffeine and on-going comparative studies with newer chemical analogues strongly point towards adenosine as the mediator of the effect. Apparently mioflazine and its analogues could well be the only family of nucleoside transport inhibitors which can reach the CNS.

In parasites or in several cultured cell lines nucleoside transport systems insensitive to NBMPR have been detected. That raised the tempting idea of protecting the host with a transport inhibitor against the cytotoxic effect of nucleoside analogues which will kill the parasite or cancer cells (see Paterson et al. 1987). Studies *in vivo* have thus far been conducted only in mice and with NBMPR (El Kouni et al. 1985). The application of other nucleoside transport inhibitors, especially of the mioflazine type (which seem much longer-acting), may prove - or disprove - the concept.

It may be concluded from the foregoing that results *in vivo* with the existing nucleoside transport inhibitors do not reveal the whole spectrum of activities which, theoretically, could be expected from the pharmacology of adenosine. A first reason could be the set-up of the experiments and incomplete understanding of the mechanisms involved: nucleoside transport inhibitors will not mimic adenosine, as do the agonists, but will act only under conditions where adenosine is produced *in vivo*. A major reason, I believe, is the bioavailability of the existing drugs which are poorly absorbed, when given orally, or very short-acting, or cannot reach their target. The development of more appropriate molecules seems mandatory. It is hard to believe that, as a third reason for the relative failure hitherto of nucleoside transport inhibition, it could be that we all have become too much excited by what the pharmacologists told us about adenosine.

REFERENCES

Reviews

R1: Berne, R.M., Rall, T.W. & Rubio, R., 1983, Regulatory Function of Adenosine (Martinus Nijhoff Publishers).

R2: Gerlach, E. & Becker, B.F., 1987, Topics and Perspectives in Adenosine Research (Springer Verlag).

R3: Paton, D.M., 1985, Methods in Pharmacology, Vol. 6: Methods Used in Adenosine Research (Plenum Press).

R4: Stefanovich, V., Rudolphi, K. & Schubert, P., 1985, Adenosine Receptors and Modulation of Cell Function (IRL Press).

Blumenthal, D.S., Hutching, G.M., Jugdutt, B.I. & Becker, L.C., 1981, Circulation, 64, 915-923.

Dawicki, D.D., Agarwal, K.C. & Parks, R.E.Jr., 1985, Biochemical Pharmacology, 34, 3965-3972.

Dendorfer, A., Lauk, S., Schaff, A. & Nees, S., 1987, In R2, p. 170-187.

Dragunow, M., 1986, Neuroscience and Behavioral Reviews, 10, 229-244.

Dresse, A., Chevolet, C., Delapierre, D., Masset, H., Weisenberger, H., Bozler, G. & Heinzel, G., 1982, European Journal of Clinical Pharmacology, 23, 229-234.

El Kouni, M.H., Knopf, P.M. & Cha, S., 1985, Biochemical Pharmacology, 34, 3921-3923.

Fredholm, B.B., Gustafsson, L.E., Hedqvist, P. & Sollevi, A., 1983, In R1, p. 479-495.

Gerlach, E., Becker, B.F. & Nees, S., 1987, In R2, p. 309-320.

Harker, L.A. & Fuster, V., 1986, Journal of the American College of Cardiology, 8, 21B-32B.

Kenakin, T.P. & Leighton, H.J., 1985, In R3, p. 213-237.

Klimt, C.R., Knatterud, G.L., Stamler, J. & Meier, P., 1986, Journal of the American College of Cardiology, 7, 251-269.

Marzilli, M., Simonetti, I., Levantesi, D., Trivella, M.G., De Nes, M., Perissinotto, A., Puntoni, R., Buzzigoli, G., Boni, C., Michelassi, C. & L'Abbate, A., 1984, American Heart Journal, 108, 276-285.

Nakagawa, Y., Gudenzi, M. and Mustafa, S.J., 1986, Pharmacology, 33, 148-156.

Newby, A.C., 1984, Trends in Biochemical Sciences, 9, 42-44.

Paterson, A.R.P., Jakobs, E.S., Ng, C.Y.C., Odegard, R.D. & Adjei, A.A., 1987, In R2, p. 89-101.

Phillis, J.W., Barraco, R.A., De Long, R.E. & Washington, D.O., 1986, Pharmacology Biochemistry & Behavior, 24, 263-270.

Picano, E., Lattanzi, F., Masini, M., Distante, A. & L'Abbate, A., 1987, American Journal of Cardiology, 59, 71-73.

Schaumlöffel, E. & Prignitz, R., 1972, Arzneimittel-Forschung, 22, 1651-1655.

Seth, S.D., Gupta, J.P. & Sing, 1986, Journal of Molecular and Cellular Cardiology, 18 (Suppl. 1), Abstr. 205.

Sollevi, A., Torsell, L., Öwall, A., Edlund, A. & Lagerkranser, M., 1987, In R2, p. 599-613.

Sono, K., Akimoto, Y., Kurahashi, K. & Fujiwara, M., 1986, Naunyn-Schmiedeberg's Archiv für Pharmakologie, 334, 145-148.

Van Belle, H., 1985, Molecular Physiology, 8, 615-630.

Van Belle, H., Wynants, J., Xhonneux, R. and Flameng, W., 1986, Cardiovascular Research, 20, 658-664.

Van Belle, H., Goossens, F. & Wynants, J., 1987, American Journal of Physiology, 252, H886-H893.

Wauquier, A., Van Belle, H., Van den Broeck, W.A.E. & Janssen, P.A.J., 1987, Psychopharmacology, 91, 434-439.

CARDIAC AND RENAL ACTIONS OF ADENOSINE ANTAGONISTS IN VIVO

Michael G Collis
Bioscience II Department, ICI Pharmaceuticals Division, Alderley Park, Macclesfield, Cheshire SK10 4TG U.K.

Introduction

Adenosine has been proposed to mediate many physiological responses by its effects on blood flow and tissue function (see Berne et al., 1983). The evidence presented in support of these proposed roles for endogenous adenosine is generally of four different types. Firstly, administration of exogenous adenosine can evoke a biological response. Secondly, other stimuli that evoke the same response can elevate the concentration of adenosine in the tissue exhibiting the response or in fluids effluxing from that tissue. Thirdly, compounds which can prevent the uptake and metabolism of adenosine can enhance the response. Fourthly, agents which destroy adenosine or block its' effects can attenuate or abolish the response. It is ideal to have positive evidence of all four types to prove a role for adenosine in the genesis of a particular response. It is the authors prejudice, however, that of these four types of experimental observation, the most compelling evidence for or against a role for adenosine in a response is obtained by the use of antagonists with selective affinity for adenosine receptors.

The demonstration that administration of exogenous adenosine can evoke a particular response, although interesting, cannot be regarded as evidence for an endogenous role for the purine, but rather as a stimulus for further investigation. The association of a response with an elevation of tissue or tissue perfusate concentrations of adenosine is also only suggestive of an association between the two phenomena. In order to make quantitative comparisons between adenosine concentrations and the amplitude of the response, it is important to measure levels of adenosine in the interstitial fluid. This concentration cannot be estimated from whole tissue samples as intra- and extra-cellular pools of the purine cannot be distinguished (Olsson et al., 1982). Estimates based on the venous adenosine concentration are also of limited relevance since the purine is rapidly taken up and metabolised by platelets, erythrocytes and vascular endothelial cells (Dresse et al., 1982; Jarvis et al., 1983; Pearson et al., 1978). The adenosine concentration in fluids bathing various tissues have been used to assess the interstitial fluid concentration of adenosine, however, it is not clear that equilibrium exists between these fluids and the interstitial fluid. Thus, the inability to rapidly measure interstitial fluid concentrations of adenosine prevents quantitative correlations with an observed tissue response.

Compounds that inhibit disposition mechanisms for adenosine have also been used to provide evidence of the purines role in a response. If such a compound enhances the amplitude of the response then an involvement of adenosine is implied. Such an observation is, however, of limited value since the artificial elevation of endogenous adenosine concentrations into the range where they evoke a response does not prove that such concentrations ever occur in the absence of the potentiating agent.

The fourth approach to the investigation of the role of adenosine is by the use degradative enzymes or receptor antagonists. This is the equivalent of the classical technique of removing an endocrine gland in order to assess the function of its secretions. Since all mammalian cells have the capability of producing adenosine, a "chemical scalpel" is required to remove it. The value of the evidence obtained by this technique is critically dependant on the efficiency and selectivity of the "chemical scalpel" that is used. The enzyme adenosine deaminase has been extensively employed to investigate the role of adenosine. There is controversy, however, concerning the ability of the enzyme to rapidly destroy sufficient adenosine in the interstitial fluid. A further complication of the use of the enzyme is that it generates inosine. Inosine is known to have weak vasodilator effects (Collis et al., 1986b), positive inotropic effects, and can potentiate the effects of adenosine by competition for the nucleoside transporter (Ngai et al., 1987). Because of these difficulties, the use of adenosine receptor antagonists is preferable to the use of adenosine deaminase in establishing a role for adenosine in the mediation of a particular response.

Until recently, the adenosine antagonists that were available (theophylline, caffeine, I.B.M.X.) had low affinity for the adenosine receptor (Fredholm and Persson, 1982) and also inhibited phosphodiesterase enzymes (Fredholm et al., 1979). This inhibitory effect on phosphodiesterase enzymes has made the results obtained with these compounds difficult to interpret, particularly in situations where activation of adenosine receptors results in an increase in cyclic AMP levels. The recent development of more potent and selective adenosine receptor antagonists has provided better pharmacological tools with which to assess the role of adenosine.

The Pharmacology of 8-phenyltheophylline

8-phenyltheophylline is a reasonably potent adenosine antagonist (pA_2 6.3–6.5, Collis et al., 1985) with minimal effects on phosphodiesterase (Smellie et al., 1979). This alkylxanthine has no antagonist effect at cardiac or ileal muscarinic receptors, at histamine H_1 receptors or at α_1, α_2, β_1 or β_2 adrenoceptors *in vitro* at a concentration of 10^{-5}M (Collis, unpublished observations). The selectivity of 8-phenyltheophylline for adenosine receptors can also be demonstrated *in vivo*. Fig. 1 shows dose-response curves to the vasodilator effect of intra-arterial adenosine, ATP and isoprenaline in the blood perfused hindlimb of the dog. An infusion of 8-phenyltheophylline (69 µg/kg/min) causes a marked rightwards displacement of the dose-response curve to adenosine but has no effect on responses evoked by ATP and isoprenaline.

Thus, our studies with 8-phenyltheophylline both *in vitro* and *in vivo* have failed to demonstrate any other pharmacological effect of the xanthine apart from a potent antagonist effect at the adenosine receptor. We have investigated the effects of 8-phenyltheophylline on some aspects of cardiac and renal function in order to determine whether endogenous adenosine has a regulatory role in these organs during physiological and pathophysiological situations.

FIGURE 1.
DOSE RESPONSE CURVES TO ATP, ADENOSINE AND ISOPRENALINE IN THE DOG PERFUSED HIND – LIMB ± 8 – PHENYLTHEOPHYLLINE (69ug/kg/min. i.v.)

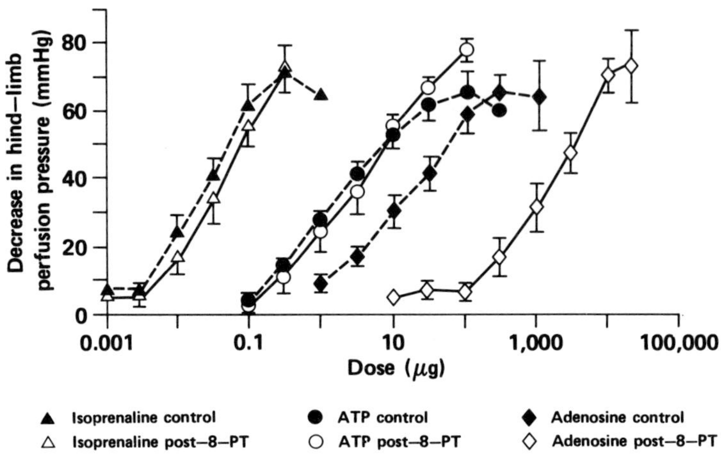

The Cardiac Actions of 8-phenyltheophylline

Adenosine has been shown to have a number of effects on cardiac function. Administration of the purine can decrease sino-atrial node rate (Collis and Saville, 1984), can cause coronary dilation (Feigl, 1983) and can antagonise the positive inotropic effects of catecholamines in the ventricle (Schrader, 1981). In order to determine which of these effects are of physiological or pathophysiological importance, we have investigated the effects of 8-phenyltheophylline on cardiac function in the anaesthetised dog before, during and after a 1 minute period of occlusion of the left anterior descending coronary artery (LAD). A dose of 5 mg/kg (i.v.) of 8-phenyltheophylline was used as we have shown that it causes a substantial blockade of responses to exogenous adenosine in the dog (Collis et al., 1984)

In control conditions, 8-phenyltheophylline had no significant effect on coronary blood flow or pressure, on left ventricular dP/dt max., on pressure rate product or on the velocity of shortening of a segment of the myocardium that was perfused by the LAD (Fig.2, Table 1). A small, but significant increase in heart rate was noted in animals treated with 8-phenyltheophylline (Table 1). Occlusion of the LAD for one minute caused cessation of blood flow, a decrease in coronary perfusion pressure (distal to the occlusion) and a marked decrease in the velocity of shortening of the segment of myocardium which was normally perfused by the LAD (Table 1). 8-phenyltheophylline had no effect on any parameter during LAD occlusion. On reperfusion, there was a marked increase in coronary artery blood flow and a recovery of shortening velocity in the reperfused myocardium. 8-phenyltheophylline significantly attenuated the hyperaemia

Table 1 Effect of 8-phenyltheophylline (8PT, 5mg/kg, i.v.) on Haemodynamic Parameters in the Anaesthetised Dog

Parameter	LAD occlusion	Vehicle	8PT
Heart Rate (b/min)	Before	110.8 ± 1.7	114.8 ± 2.4 (1)
	During	109.9 ± 1.5	113.3 ± 2.4
	After	110.4 ± 2.2	114.2 ± 2.3 (1)
Pressure Rate Product (mHg/min)	Before	13.0 ± 0.9	13.8 ± 1.1
	During	12.8 ± 0.8	13.4 ± 1.0
	After	13.1 ± 0.8	13.9 ± 0.8
LAD pressure (mmHg)	Before	87.3 + 3.5	85.5 ± 2.9
	During	24.6 ± 1.2	23.2 ± 0.7
	After	76.3 ± 1.7	78.1 ± 2.3
LV dP/dt max (mmHg/sec)	Before	2289 ± 221	2489 ± 382
	During	2048 ± 204	2186 + 309
	After	2435 + 259	2531 ± 332
$V^{1/3rd}$* (mm/cm/sec)	Before	9.1 ± 1.8	9.5 ± 2.4
	During	1.9 ± 1.7	3.5 ± 2.8
	After	6.2 ± 1.3	6.6 ± 1.5

* Velocity of shortening during the first third of systole in an area of left ventricle perfused by the left anterior descending coronary artery (LAD)

Control values (before) were taken over a 2 minute period, values during and after LAD occlusion were taken over the one minute period of occlusion and over the minute following occlusion. (1) $P<0.05$ relative to vehicle, n=4.

following the release of the LAD occlusion (Fig.2). The alkylxanthine had no significant effect on the other haemodynamic parameters on reperfusion with the exception of a small increase in heart rate (Table 1).

Figure 2.

Effect of 8 – Phenyltheophylline (5mg/kg, i.v.) on Left Anterior Descending Coronary Flow, Before and After Occlusion of the Artery for One Minute

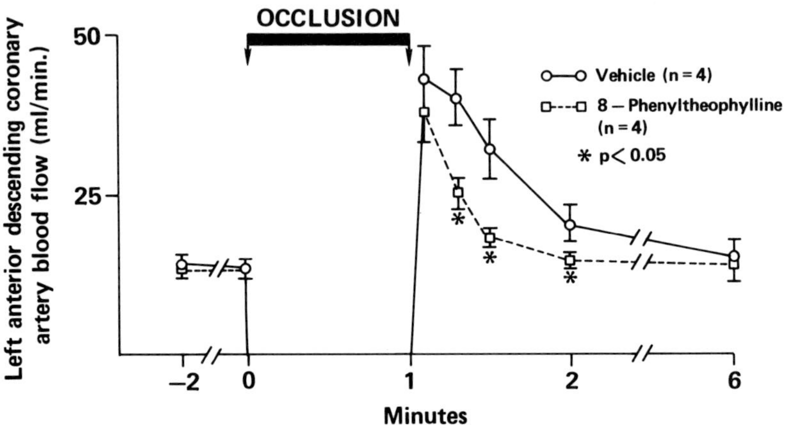

The results obtained in the anaesthetised dog indicate that adenosine is unlikely to regulate normal coronary vascular tone or cardiac contractile function, with the exception of a small negative effect on heart rate. Following a short period of ischaemia, however, adenosine may be one of the mediators of reactive hyperaemia since 8–phenyltheophylline attenuates this response. A similar effect of 8–phenyltheophylline has been reported in the conscious dog (Bache et al., 1985). The lack of either a global or a regional positive inotropic effect of 8–phenyltheophylline, even after ischaemia when adenosine levels are enhanced (Feigl, 1983), argues against a role for the purine in regulating myocardial contractility in the dog. The lack of effect of 8–phenyltheophylline cannot have been due to a low level of adrenergic sympathetic drive to the heart, since a β–adrenoceptor blocking agent will cause a marked negative inotropic effect in this preparation (Collis et al., 1984).

The Renal Actions of 8–phenyltheophylline

The renal effects of exogenous adenosine include transient renal vasoconstriction and a sustained decrease in G.F.R. (Osswald et al., 1978). The excretion of sodium and water are also decreased by the purine (Spielman and Thompson, 1982). If these effects are of physiological relevance, than 8–phenyltheophylline should increase G.F.R. and enhance sodium and water excretion.

We have examined the effects of 8–phenyltheophylline on renal function in the rat. The effects of this adenosine antagonist have been compared with those of another alkylxanthine, enprofylline, which has a low affinity for adenosine receptors (Collis et al., 1986a). The dose of 8–phenyltheophylline (10 mg/kg, i.p.) which was used in this study has been shown to antagonise significantly the bradycardic effect of intravenous adenosine for longer than 3 hours in the conscious rat. The same dose of enprofylline, however, had no significant effect on adenosine induced bradycardia (Collis et al., 1986a). In order to evaluate excretory function, rats were given 40 ml/kg of saline (p.o.) and their urine was collected for 6 hours. 8–phenyltheophylline evoked a significant increase in urine volume and in the excretion of sodium and chloride when compared to animals treated with the vehicle (Fig.3). Enprofylline, however, was without a significant effect on these parameters (Fig.3).

The results of this study therefore imply that endogenous adenosine exerts an influence on renal function in the normal rat. The question then arises of whether this is due to an effect on G.F.R. In further studies, H^3 inulin clearance was assessed in the anaesthetised rat as an index of G.F.R. 8–phenyltheophylline (10 mg/kg i.p.) had no significant effect on inulin clearance when compared with vehicle–treated control animals (1.07 ± 0.01 and 0.99 ± 0.1 ml/min/100g body weight, respectively, Bowmer et al., 1986). Thus, the diuretic and saliuretic effect of 8–phenyltheophylline in the normal rat is not associated with an increase in G.F.R.

Figure 3.
Effect of 8 – Phenyltheophylline and Enprofylline on Excretory Parameters in the Rat

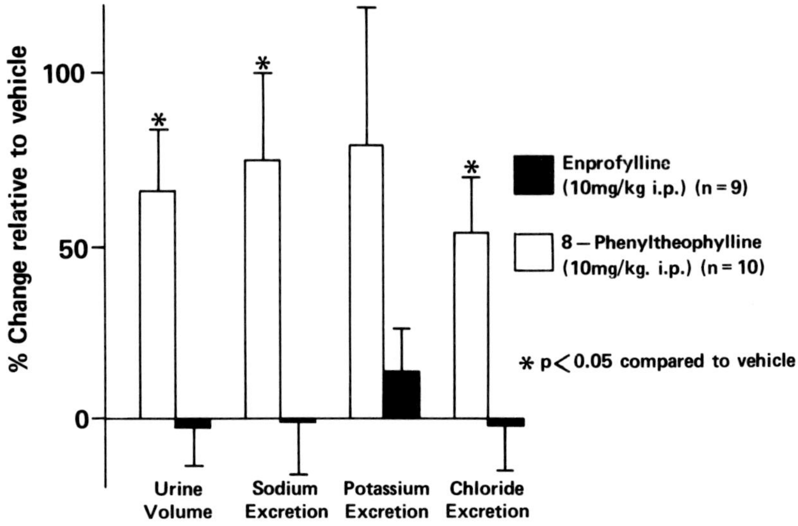

Adenosine has been shown to stimulate sodium transport in toad kidney epithelial cells (Lang et al., 1985). It is therefore possible that the diuretic effect of 8-phenyltheophylline in the rat could be due to blockade of the effects of endogenous adenosine on tubular ion transport.

The effects of 8-phenyltheophylline in the normal rat provide evidence for a role of adenosine in the physiological regulation of renal function, but not via an action on G.F.R. There may, however, be pathophysiological situations in which adenosine levels increase sufficiently to have an effect on G.F.R. The intramuscular injection of glycerol induces a state of acute renal failure in the rat. This is associated with a transient reduction in renal blood flow (Hsu et al., 1977) and a major reduction in renal concentrations of adenine nucleotides (Trifillis et al., 1981). Both renal ischaemia and a reduction in nucleotide levels have been associated with an elevation of renal adenosine concentrations (Miller et al., 1978). In rats with glycerol–induced acute renal failure, 8–phenyltheophylline enhanced inulin clearance from 0.35 ± 0.04 to 0.60 + 0.09 ml/min/100g body weight (Bowmer et al., 1986). This apparent improvement in an index of G.F.R. is not a consequence of a systemic haemodynamic effect of 8–phenyltheophylline as blood pressure was not altered by the alkylxanthine. The most likely explanation of the results is that intrarenal adenosine reduces G.F.R. in acute renal failure and that this effect is antagonised by 8–phenyltheophylline. In addition to its beneficial effects on G.F.R., 8–phenyltheophylline also reduced plasma urea and creatinine levels, reduced kidney weight and the morphological damage associated with this form of renal failure and reduced the mortality rate (Bowmer et al., 1986). It is not known whether these effects of 8–phenyltheophylline are secondary to the improvement in G.F.R. or to actions at other sites in the kidney.

Conclusions

The potent and selective adenosine receptor antagonist, 8–phenyltheophylline, has been used to investigate the role of adenosine in some aspects of cardiac and renal function. The results of these studies support a role for endogenous adenosine in post–ischaemic coronary hyperaemia, but not in the control of cardiac inotropic state. At least two effects of endogenous adenosine have been revealed with respect to renal function. In the normal animal, adenosine may be involved in regulating tubular function. In the more extreme situation of acute renal failure, the purine also appears to be important in regulating G.F.R. The further use of 8–phenyltheophylline and of other selective adenosine antagonists should help to reveal further roles for adenosine in physiological and pathophysiological situations.

References

Bache, R.J., Schwartz, J.S. and Dia, X–Z. 1985. Journal of the American College of Cardiology, 5, 444.

Berne, R.M., Rall, T.W. and Rubio, R. 1983. Regulatory Function of Adenosine (Martinus Nijhoff, The Hague).

Bowmer, C.J., Collis, M.G. and Yates, M.S. 1986. British Journal of Pharmacology, 88, 205.

Collis, M.G., Baxter, G.S. and Keddie, J.R. 1986a). Journal of Pharmacy and Pharmacology, 38, 850.
Collis, M.G., Keddie, J.R. and Torr, S.R. 1984. British Journal of Pharmacology, 81, 401.
Collis, M.G., Palmer, D.B. and Baxter, G.S., 1986b European Journal of Pharmacology, 121, 141.
Collis, M.G., Palmer, D.B. and Saville, V.L. 1985. Journal of Pharmacy and Pharmacology, 37, 278.
Collis, M.G. and Saville, V.L. 1984, British Journal of Pharmacology, 83, 413P.
Dresse, A., Chevolet, C., Delapierre, D., Masset,
Weisenberger, H., Bolzer, G. and Heinzel, H. 1982, European Journal of Clinical Pharmacology, 23, 229.
Feigl, E.O. 1983, Physiological Reviews, 63, 1.
Fredholm, B.B., Brodin, K. and Strandberg, K. 1979, Acta Pharmacologica et Toxicologica, 45, 336.
Fredholm, B.B. and Persson, C.G.A. 1982, European Journal of Pharmacology, 81, 673.
Hsu, C.H. Kurtz, T.W. and Waldinger, T.P. 1977, Circulation Research, 40, 178.
Jarvis, S.M., Hammond, J.R., Paterson, A.R.P. and Clanachan, A.S. 1983. Biochemical Journal, 210, 457.
Lang, M.A., Preston, A.S., Handler, J.S. and Forrest, J.N. 1985, American Journal of Physiology, 249, C330.
Miller, W.L., Thomas, R.A., Berne, R.M. and Rubio, R. 1978, Circulation Research, 43, 390.
Ngai, A.C., Ibayashi, S., Mero, J.R. and Winn, H.R. 1987, Federation Proceedings, 46, 800.
Olsson, R.A., Saito, D. and Steinhart, C.R. 1982. Circulation Research, 50, 617.
Osswald, H., Spielman, W.S., Knox, F.G. 1978, Circulation Research, 43, 465.
Pearson, J.D., Carleton, J.S., Hutchings, A. and Gordon, J.L. 1978, Biochemical Journal, 170,265.
Schrader, J. 1981. In Purinergic Receptors edited by G Burnstock, (Chapman Hall, London) p119.
Smellie, F.W., Davis, C.W., Daly, J.W. and Wells, J.N. 1979. Life Science, 24, 2475.
Spielman, W.S. and Thompson, C.I. 1982, American Journal of Physiology, 242, F423.
Trifillis, A.L., Kahng, M.W. and Trump, B.F. 1981, Experimental and Molecular Pathology, 35, 1.

ACKNOWLEDGEMENT

The experiments on cardiac function described in this article were performed in collaboration with Dr W Rouse.

MECHANISM OF ACTION OF METHYLXANTHINES

THE SITE AND MECHANISM OF THE RELAXANT EFFECTS OF AMINOPHYLLINE AND OTHER METHYLXANTHINES IN ISOLATED AIRWAYS SMOOTH MUSCLE

R C Small, R W Foster, J P Boyle & J M Davies

Smooth Muscle Research Group
Department of Physiological Sciences
University of Manchester, Manchester M13 9PT, UK

Relaxation of airways smooth muscle: the site of action of methylxanthines

For many years it has been assumed that the methylxanthines exert their relaxant effects in airways smooth muscle by crossing the plasmalemma and inhibiting intracellular phosphodiesterase enzymes. However, this assumption has recently been challenged. 1,3,7,9-tetramethylxanthinium is a quaternized xanthine derivative which, though less potent than theophylline, retains the ability to relax airways smooth muscle in vitro. In view of this, and since permanently charged xanthines should not readily cross the plasmalemma, it has been argued (Persson 1985a,b) that the relaxant activity of methylxanthines may be mediated at a receptor located on the external surface of the cell.

We have recently been able to confirm the relaxant activity of 1,3,7,9-tetramethylxanthinium (8-32 mM) tested against the spontaneous tone of guinea-pig isolated trachealis as observed at 37°C. However, when similar concentrations of the xanthinium are tested on trachealis treated with 2.8 μM indomethacin and maintained at 12°C, spasmogenic effects are observed. Spasm is evoked by caffeine, enprofylline, theobromine and theophylline under similar conditions (Boyle et al. 1987).

The spasmogenic effects of caffeine in airways smooth muscle can be observed in a Ca^{2+}-free, EGTA-containing medium (Ito & Itoh 1984a; Boyle et al. 1987) and represent the release of Ca^{2+} from intracellular sites of sequestration (Ito & Itoh 1984a,b). Since caffeine has an action of this kind in saponin-permeablized smooth muscle (Meisheri et al. 1985) it seems probable that the Ca^{2+}-releasing effect of the methylxanthines represents their direct action at the sites of Ca^{2+} sequestration within the cell. Accordingly, the spasmogenic activity of 1,3,7,9-tetramethylxanthinium may be an index of its ability to cross the plasma membrane. The usefulness of 1,3,7,9-tetramethylxanthinium as an analytical tool is therefore questionable and localisation of the site at which the relaxant actions of the methylxanthines are mediated remains an outstanding problem.

Phosphodiesterase inhibition induced by methylxanthines: its role in the relaxant process

As mentioned above, inhibition of intracellular phosphodiesterase is commonly assumed to underlie the relaxant effects of methylxanthines in isolated airways smooth muscle. Inhibition of this enzyme is envisaged to cause the intracellular accumulation of cAMP. The activation of cAMP-dependent protein kinases may then stimulate a variety of processes which reduce the cytosolic concentration of free Ca^{2+} or reduce the Ca^{2+}-sensitivity of the intracellular contractile machinery (Rodger 1986).

What evidence has been obtained in airways smooth muscle to suggest that cAMP phosphodiesterase inhibition plays a key role in the relaxant actions of the methylxanthines ? Theophylline and other methylxanthines have been shown to inhibit cAMP phosphodiesterase derived from trachealis smooth muscle of the ox, guinea-pig and dog (Lohmann et al. 1977; Newman et al. 1978; Polson et al. 1979,1982) or from lung tissue of humans and guinea-pigs (Bergstrand 1980; Persson 1985b). Enzyme inhibition is observed at concentrations of methylxanthines below those producing maximal relaxant effects (Polson et al. 1979, 1982). Moreover a good correlation exists between the rank order of potency of methylxanthines in relaxing trachealis and their rank order of potency in inhibiting cAMP phosphodiesterase (Polson et al. 1982; Persson 1985b).

While most authors agree that methylxanthines can inhibit cAMP phosphodiesterase derived from respiratory tract tissue, findings with regard to the ability of methylxanthines to modify tissue cAMP content are less consistent. In guinea-pig or bovine trachealis not exposed to exogenous spasmogens, theophylline (1-10 mM) was observed to increase the tissue content of cAMP (Katsuki & Murad 1977; Kolbeck et al. 1979). However, Kolbeck et al. (1979) found no theophylline-induced change in the cAMP content of canine trachealis and Lohmann et al. (1977) were unable to confirm that theophylline increased the cAMP content of bovine trachealis. The latter workers also failed to observe an increase in the cAMP content of bovine trachealis during theophylline-induced suppression of spasm evoked by carbachol.

The failure of methylxanthines consistently to raise the

cAMP content of airways smooth muscle has led to critical re-appraisal (Kolbeck et al. 1979; Persson 1985b) of the role of cAMP in mediating methylxanthine-induced relaxation of airways smooth muscle. However, it is possible that very small changes in the cellular content of cAMP mediate the relaxant action or that the relevant increase in cAMP concentration occurs in a limited subcellular compartment (Lohmann et al. 1977). Such changes would not be recognised in experiments involving measurements of the total tissue content of cAMP.

The above mentioned re-appraisal of the role of cAMP in mediating methylxanthine-induced relaxation has also been stimulated by the observation that other agents with the ability to inhibit cAMP phosphodiesterase are ineffective in relaxing isolated airways smooth muscle. Disodium cromoglycate, for example, is without relaxant effect on human isolated airways smooth muscle (Cox 1967) yet inhibits cAMP phosphodiesterase obtained from human lung (Bergstrand 1980). The role of phosphodiesterase inhibition and cAMP accumulation in mediating the relaxant effects of methylxanthines, therefore, has yet to be established.

Membrane potential changes evoked by methylxanthines: their underlying mechanisms and their role in the relaxant process

The resting membrane potential of guinea-pig trachealis is approximately -50 mV and most cells exhibit oscillations of membrane potential known as slow waves. The slow waves represent an aborted form of the regenerative action potential and are prevented from attaining full spike-like configuration by the activation of voltage-dependent K^+-channels (Small & Foster 1987, 1988).

Aminophylline and theophylline have relaxant effects in this tissue which are associated with slow wave suppression and hyperpolarisation (Allen et al. 1986b; Honda et al. 1986; Figure 1). These electrical changes can be mimicked by the adenylate cyclase activator, forskolin, and by dibutyryl-cAMP (Honda et al. 1986) and hence may result from the intracellular accumulation of cAMP.

Membrane conductance changes have not been measured during methylxanthine-induced hyperpolarisation of airways smooth muscle. However, two findings suggest that the hyperpolarisation represents the opening of K^+-channels in the cell membrane. Firstly, aminophylline-induced hyperpolarisation approached the calculated (-77 mV; Kirkpatrick 1981) K^+-equilibrium potential of trachealis smooth muscle. Secondly, the K^+-channel inhibitors, tetraethylammonium (TEA) and procaine, were each able to reduce the aminophylline-induced hyperpolarisation (Allen et al. 1986b).

Figure 1. The effects of aminophylline (1 mM) on the electrical and mechanical activity of guinea-pig isolated trachealis. In each case the upper trace represents membrane potential and the lower trace the mechanical activity of a contiguous segment of trachea. All electrical recordings are taken from the same cell. Activity was recorded before, and 1.5, 3 and 4.5 min after addition of aminophylline (1 mM) at the arrow. Note the abolition of slow waves and the marked hyperpolarisation which accompany the aminophylline-induced relaxation. (Adapted from Allen et al. 1986b with permission).

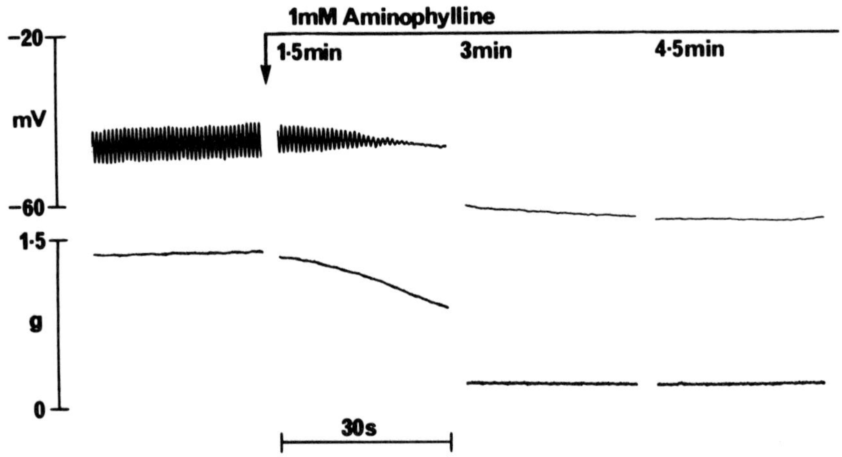

Despite causing hyperpolarisation attributable to K^+-channel opening, aminophylline does not promote $^{86}Rb^+$ efflux from $^{86}Rb^+$-preloaded, muscle-rich tracheal strips (Allen et al. 1986b). This suggests that the K^+-channels opened by the methylxanthines are similar to those opened by isoprenaline (Allen et al. 1985, 1986c) but distinct from the $^{86}Rb^+$-permeable K^+-channels opened by nicorandil or BRL34915 (Allen et al. 1986a,c).

Methylxanthine-induced hyperpolarisation might be envisaged to play an important role in the process by which relaxation is achieved. A more negative resting membrane potential would certainly reduce the likelihood of the opening of voltage-operated Ca^{2+}-channels in the cell membrane and hence reduce Ca^{2+} influx. Trachealis slow waves are suppressed by relatively small excursions from the normal value of the resting membrane potential (Kirkpatrick 1981) and the depolarising phase of the slow waves is Ca^{2+}-carried (Small & Foster 1987, 1988). Methylxanthine-induced slow wave suppression therefore may depend on the drug-induced hyperpolarisation and produces a reduction in Ca^{2+} influx.

However, several observations suggest that inhibition of Ca^{2+} influx by this mechanism is relatively unimportant for expression of the relaxant effects of the methylxanthines: (1) Low concentrations of aminophylline produce relaxation in the absence of membrane potential changes (2) Aminophylline retains relaxant activity in a K^{+}-rich (120 mM) medium where hyperpolarisation cannot be achieved by K^{+}-channel opening because the resting membrane potential and the K^{+}-equilibrium potential are virtually co-incident (3) The K^{+}-channel inhibitors TEA and procaine can significantly reduce aminophylline-induced hyperpolarisation without reducing its relaxant effects (Allen et al. 1986b).

We therefore conclude that the membrane potential changes evoked by methylxanthines represent a by-product (which may play a minor supportive role) of, rather than an essential link in, the process by which relaxation of airways smooth muscle is achieved.

Influence of methylxanthines on the role of membrane phospholipids in excitation-contraction coupling

In airways smooth muscle there is substantial evidence (Small & Foster 1986, 1988) that spasmogens which act at structurally specific receptors on the cell surface (eg agonists at muscarinic cholinoceptors, agonists at histamine H_1 receptors) evoke contraction primarily by releasing Ca^{2+} from intracellular stores. The belief is growing that the mechanism linking receptor activation to Ca^{2+} release from the intracellular stores involves the hydrolysis of phosphatidyl-inositol phosphates in the plasmalemma to yield inositol phosphates and diacylglycerols. Acetylcholine, carbachol and histamine have each been shown to stimulate phosphatidyl-inositol phosphate breakdown or inositol phosphate accumulation in airways smooth muscle (Baron et al. 1984; Hashimoto et al. 1985; Barnes et al. 1986; Grandordy et al. 1986) and inositol trisphosphate has been shown to stimulate the release of Ca^{2+} from intracellular storage sites in saponin-treated canine trachealis (Hashimoto et al. 1985).

Methylxanthines can suppress the tone developed spontaneously by airways smooth muscle and the elevated tone induced by challenge with an agonist at muscarinic cholinoceptors or with histamine. However, when tone is elevated by a spasmogen (particularly in the case of agonists at muscarinic cholinoceptors) the concentration-effect curve of methylxanthine may be shifted to the right (Karlsson & Persson 1981; Allen et al. 1986b). The ability of methylxanthines to relax airways smooth muscle pre-contracted with spasmogens which activate cell surface receptors is unlikely to be attributable to interference with the production of inositol phosphate second messengers for, in bovine trachealis treated with Li^{+}, theophylline (100 μM) did not depress the accumulation of inositol monophosphate induced by carbachol, histamine or 5HT (Grandordy & Barnes 1986; Grandordy et al. 1987). However, a possibility as yet untested, is that the methylxanthines interfere with the actions of the inositol phosphate second messengers at the level of the intracellular Ca^{2+} storage sites.

The influence of methylxanthines on the intracellular effects and disposition of Ca^{2+}

As stated above, agonists at muscarinic or histamine H_1 receptors evoke spasm of airways smooth muscle principally by releasing Ca^{2+} from intracellular sites of sequestration. In contrast potassium chloride (KCl) principally acts to promote the cellular influx of Ca^{2+} through voltage-operated Ca^{2+} channels (Small & Foster 1986, 1988). Methylxanthines can suppress both the spasm evoked by KCl and the spasm evoked by agonists at muscarinic cholinoceptors or histamine H_1 receptors (Karlsson & Persson 1981; Allen et al. 1986b, Nielsen-Kudsk et al. 1986). This may suggest that the effects, or the amplitude of, the increase in cytosolic free Ca^{2+} concentration induced by spasmogens can be reduced by methylxanthines in a manner which is independent of the Ca^{2+} source (extracellular fluid or intracellular stores).

The possibility that methylxanthines might act to reduce the responsiveness of the intracellular contractile machinery to cytosolic free Ca^{2+} has been tested in trachealis chemically-skinned of its plasmalemmal membranes. The detergent Triton X-100 is an effective tool for destroying the plasmalemma (Sparrow et al. 1984; Cortijo et al. 1987) and the skinned fibres prepared using this agent develop tension when exposed to free Ca^{2+} in the concentration range 0.2-20 μM. Neither aminophylline (Allen et al. 1986b) nor isobutylmethylxanthine (Bryson & Rodger 1987) was able to depress the maximum spasmogenic response to Ca^{2+} in Triton X-100 skinned trachealis. This suggests that methylxanthines do not directly reduce the Ca^{2+} responsiveness of the intracellular contractile machinery. However, in the intact cell, the possibility remains that the methylxanthines can reduce the potency or effectiveness of Ca^{2+} by indirect mechanisms (eg protein kinase activation) which may depend on the presence of an intact plasmalemma and/or cytosol.

What evidence is there that the methylxanthines might exert relaxant effects by altering the intracellular disposition of Ca^{2+}? As discussed above, the spasmogenic effects of caffeine in feline trachealis have been attributed to an action of the methylxanthine in releasing Ca^{2+} from intracellular sites of sequestration (Ito & Itoh 1984a,b). Spasmogenic effects of caffeine, enprofylline, theobromine and theophylline are also seen in guinea-pig trachealis treated with 2.8 μM indomethacin and maintained at 12°C (Boyle et al. 1987). This suggests that the ability to release Ca^{2+} from intracellular stores is a property shared by various methylxanthines.

The release of Ca^{2+} from the stores might be regarded as a process which opposes the induction of relaxation. However, Ca^{2+} release could result in depletion of the stores and, in feline trachealis bathed by a Ca^{2+}-free, EGTA-containing medium, Ito & Itoh (1984a) obtained evidence that 10 mM caffeine can prevent acetylcholine-induced spasm by such a mechanism. It seems unlikely that depletion of intracellular Ca^{2+} stores can explain the relaxant actions of methylxanthines as observed in a medium containing physiological concentrations of Ca^{2+}. In the presence of Ca^{2+} (2.6 mM) enprofylline and theophylline can evoke relaxation at concentrations well below those required to evoke spasm attributable to Ca^{2+} release from intracellular stores (Boyle et al. 1987). Furthermore, depletion of Ca^{2+} stores would not explain the relaxant effects of methylxanthines when tested against spasmogens (eg KCl) whose actions result from Ca^{2+} influx rather than release from intracellular stores.

If, as suggested by Ito & Itoh (1984b) the intracellular stores are normally saturated with Ca^{2+} and if it can be assumed that KCl does not release Ca^{2+} from the stores, then the ability of methylxanthines to suppress KCl-induced spasm cannot be explained in terms of the methylxanthines promoting the intracellular sequestration of Ca^{2+}. We are left with the possibility either that the methylxanthines (a) reduce the Ca^{2+} sensitivity or responsiveness of the contractile machinery by the indirect mechanisms described above or (b) reduce the cytosolic concentration of free Ca^{2+} by promoting its extrusion from the cell. Both these mechanisms await the test of experiment.

CONCLUSIONS

1 New evidence offered on the activity of a quaternary tetramethylxanthinium suggests that whether the action of methylxanthines in causing relaxation of airways smooth muscle is exerted extra- or intracellularly is still an open question.

2 The circumstantial evidence for involvement of inhibition of cAMP-phosphodiesterase seems good but the necessary causal increase in the size of the messenger cAMP signal has not been measured convincingly. It seems wise to continue to examine other possible mechanisms of action.

3 Membrane hyperpolarisation and suppression of electrical slow waves occur. The hyperpolarisation may result from cAMP accumulation and seems to be mediated by the opening of K^+ channels. However the membrane potential changes are not essential to the process whereby relaxation is produced.

4 Methylxanthines do not change the size, but may reduce the effectiveness, of the messenger IP_3 signal in releasing Ca^{2+} from intracellular storage sites.

5 They do not directly reduce the effectiveness of Ca^{2+} on the contractile machinery but may do so by indirect means. They may reduce the size of the messenger Ca^{2+} signal by promoting Ca^{2+} extrusion from the cytosol.

ACKNOWLEDGEMENTS

Supported by the Asthma Research Council, the British Lung Foundation, Napp Laboratories Ltd, the North Western Regional Health Authority and the Wellcome Trust.

REFERENCES

Allen, S.L., Beech, D.J., Foster, R.W., Morgan, G.P. & Small,R.C., 1985, British Journal of Pharmacology, 86, 843.

Allen, S.L., Boyle, J.P., Cortijo, J., Foster, R.W., Morgan,G.P. & Small, R.C., 1986a, British Journal of Pharmacology, 89, 395.

Allen, S.L., Cortijo, J., Foster, R.W., Small, R.C.& Weston,A.H., 1986b, British Journal of Pharmacology, 88, 473.

Allen, S.L., Foster, R.W., Morgan, G.P. & Small, 1986c, British Journal of Pharmacology, 87, 117.

Barnes, P.J., Cuss, F.M. & Grandordy, B.M., 1986, British Journal of Pharmacology, 87, 65P.

Baron, C.B., Cunningham, M., Strauss, J.F. & Coburn, R.F., 1984, Proceedings of the National Academy of Science, 81, 6899.

Bergstrand, H., 1980, European Journal of Respiratory Disease, 61, Suppl. 109, 37.

Boyle, J.P., Cortijo, J., Davies, J.M., Foster, R.W. & Small, R.C., 1987, unpublished observations.

Bryson, S.E. & Rodger, I.W., 1987, British Journal of Pharmacology, (in the press).

Cortijo, J., Dixon, J.S., Foster, R.W. & Small, R.C., 1987, Journal of Pharmacological Methods, (in the press).

Cox, J.S.G., 1967, Nature, 216, 1328.

Grandordy, B.M. & Barnes, P.J., 1986, American Reviews of Respiratory Disease, 133(4), A179.

Grandordy, B.M., Cuss, F.M. & Barnes, P.J., 1987, Life Sciences (in the press).

Grandordy, B.M., Cuss, F.M. Sampson, A.S., Palmer, J.B. & Barnes, P.J., 1986, Journal of Pharmacology and Experimental Therapeutics, 238, 273.

Hashimoto, T., Hirata, M. & Ito, Y., 1985, British Journal of Pharmacology, 86, 191.

Honda, K., Satake, T., Takagi, K. & Tomita, T., 1986, British Journal of Pharmacology, 87, 665.

Ito, Y. & Itoh, T., 1984a, British Journal of Pharmacology, 83, 667.

Ito, Y. & Itoh, T., 1984b, British Journal of Pharmacology, 83, 677.

Karlsson, J.A. & Persson, C.G.A., 1981, British Journal of Pharmacology, 74, 73.

Katsuki, S. & Murad, F., 1977, Molecular Pharmacology, 13, 330.

Kirkpatrick, C.T., 1981 Tracheobronchial smooth muscle. In Smooth Muscle: an Assessment of Current Knowledge, edited by E. Bulbring, A.F. Brading, A.W. Jones & T. Tomita (Edward Arnold: London), p. 385.

Kolbeck, R.C., Speir, W.A., Carrier, G.O. & Bransome, E.D., 1979, Lung, 156, 173.

Lohmann, S.M., Miech, R.P. & Butcher, F.R., 1977, Biochemica et Biophysica Acta, 499, 238.

Meisheri, K.D., Ruegg, J.C. & Paul, R.J., 1985 Studies on Skinned Fiber Preparations. In Calcium and Contractility: Smooth Muscle, edited by A.K. Grover & E.E. Daniel (Humana Press), p. 191.

Newman, D.J., Colella, D.F., Spainhour, C.B., Brann, E.G., Zabko-Potapovich, B. & Wardell, J.R., 1978, Biochemical Pharmacology, 27, 729.

Nielsen-Kudsk, J.E., Karlsson, J.A. & Persson, C.G.A., 1986 European Journal of Pharmacology, 128, 33.

Persson, C.G.A., 1985a Subdivision of xanthines. In Anti-asthma xanthines and adenosine, edited by K.E. Andersson & C.G.A. Persson (Excerpta Medica), p. 23.

Persson, C.G.A., 1985b Experimental lung actions of xanthines. In Anti-asthma xanthines and adenosine, edited by K.E. Andersson & C.G.A. Persson (Excerpta Medica), p. 61.

Polson, J.B., Krzanowski, J.J., Anderson, W.H., Fitzpatrick, D.F., Hwang, D.P.C., Szentivanyi, A., 1979, Biochemical Pharmacology, 28, 1391.

Polson, J.B., Krzanowski, J.J. & Szentivanyi, A., 1982, Biochemical Pharmacology, 31, 3403.

Rodger, I.W., 1986 Calcium Ions and Contraction of Airways Smooth Muscle. In Asthma: Clinical Pharmacology and Therapeutic Progress, edited by A.B. Kay (Blackwell Scientific), p. 114.

Small, R.C. & Foster, R.W., 1986 Airways Smooth Muscle: An Overview of Morphology, Electrophysiology and Aspects of the Pharmacology of Contraction and Relaxation. In Asthma: Clinical Pharmacology and Therapeutic Progress, edited by A.B. Kay (Blackwell Scientific), p. 101.

Small, R.C. & Foster, R.W., 1987, American Reviews of Respiratory Disease, (in the press).

Small, R.C. & Foster, R.W., 1988 Electrophysiology of the Airways Smooth Muscle Cell. In Asthma: Basic Mechanisms and Clinical Management, edited by P.J. Barnes, I.W. Rodger & N.C. Thompson (Academic Press: London), (in the press).

Sparrow, M.P., Pfitzer, G., Gagelmann, M. & Ruegg, J.C., 1984, American Journal of Physiology, 246, C308.

ABSTRACTS

ABSTRACTS

Modulation by purines and Ca^{2+}-antagonists of cholinergic neurotransmission in guinea-pig ileum
T. Katsuragi, K. Shirakabe, K. Miyamoto and T. Furukawa

290 Adenosine surge at birth
H. Lagercrantz, L. Irestedt, M. Runold and A. Sollevi

Adenosine and neonatal apneoa
H. Lagercrantz, M. Runold and B. Fredholm

291 Role of adenosine in regulation of mesenteric arterial blood flow
W.W. Lautt, D.J. Legare, L.A. D'Aleo, J.E. McQuaker and L.K. Lockhart

Characterization of neural and cardiac adenosine receptors by agonist and antagonist radioligands
E. Leung, K.A. Jacobson and R.D. Green

292 The role of adenosine in synaptic depression evoked by a brief period of hypoxia in the isolated rat spinal cord
H.G.E. Lloyd, I. Spence and G.A.R. Johnston

Comparison of the cardiac effects of adenosine (ADO) analogs in rat heart and their receptor binding activity in rat brain
H.H. Oei, G.R. Ghai, H.C. Zoganas, G.A. Stone, L. Sanathanan, M. Williams and M.B. Zimmerman

293 Comparative effects of 5'-N-ethylcarboxamido-adenosine (NECA) and of L-phenylisopropyl-adenosine (L-PIA) on isolated guinea-pig heart atria
M. Prostran and V.M. Varagić

Purinergic receptors in the brainstem mediate hypotension and bradycardia
D. Robertson, C.-J. Tseng, M. Appalsamy and I. Biaggioni

294 Distribution of adenosine receptors in the renal cortex
W. Schütz, C. Nanoff and M. Freissmuth

DESENSITIZATION OF ADENYLATE CYCLASE TO ADENOSINE IN VASCULAR SMOOTH MUSCLE CELLS FROM RAT AORTA.

M.B. Anand-Srivastava, M. Cantin, M. Ballak and S. Picard
Clinical Research Institute of Montreal, 110 Pine Ave West, Montréal, Quebec, Canada H2W 1R7

We have previously shown that adenylate cyclase present in rat aorta vascular smooth muscle cells can be stimulated by adenosine, its analogs and other agonists (Anand-Srivastava et al. BBRC 108, 213-219, 1982). In the present studies, we have examined the effect of pre-exposure of aorta vascular smooth muscle cells to N-ethylcarboxamide adenosine (NECA) on adenylate cyclase activity, stimulated by NECA and other agonists. The vascular smooth muscle cells, when exposed to NECA, resulted in a concentration and time dependent loss of NECA-stimulated adenylate cyclase activity. NECA, stimulated adenylate cyclase activity by about 120% in control cells, which was decreased to 20% in cells pretreated with 50 μM NECA for 30 minutes at 37°C. However, GTP-, isoproterenol-, and forskolin-sensitive adenylate cyclase activities were not affected by such treatment, suggesting that NECA treatment of the cells resulted in homologous desensitization. Similar desensitization of NECA-stimulated adenylate cyclase activity was also observed, when the cells were pretreated with 3-isobutyl-1-methylxanthine, a competitive antagonist of adenosine receptor. When NECA-treated cells were washed free of agonist, the desensitized state was reversed and the cells regained full responsiveness to NECA-stimulation of adenylate cyclase.

CARDIORESPIRATORY EFFECTS OF ADENOSINE AND CYCLIC AMP FOLLOWING MICROINJECTIONS INTO THE NUCLEUS TRACTUS SOLITARIUS OF RATS

R. Barraco, E. Schoener, P. Polasek and C. Janusz
Departments of Physiology and Pharmacology, Wayne State University Medical School, Detroit, MI U.S.A.

The Nucleus Tractus Solitarius (NTS) is a major relay nucleus for afferent input from cardiopulmonary receptors and thereby plays a critical role in brainstem regulation of cardiovascular and respiratory function. Adult male Sprague-Dawley rats were anesthetized with urethane and a limited craniotomy was conducted to expose the caudal brainstem. Mean arterial blood pressure (P_A), heart rate (HR), respiratory frequency (f) and tidal volume (T_m) were recorded following microinjections of adenosine (ADO) and 8-bromo-cyclic AMP (Br-cAMP) into the NTS. At higher doses, ADO produced decreases in P_A, HR, f and T_m while at lower doses ADO produced pressor responses. These findings may reflect an A_1/A_2 receptor-mediated response difference. To examine this, Br-cAMP was microinjected into the same site. Br-cAMP produced dose-related decreases in P_A, HR, f and T_m. These data show that ADO may play a neuromodulatory role in the NTS via actions at extracellular receptors.

DIDIOXYADENOSINE DISSOCIATES ADENOSINE AND NECA INDUCED CORONARY FLOW RESPONSE FROM ENDOTHELIAL ADENYLATE CYCLASE ACTIVATION

B.F. Becker, W.H. Newman*, S. Nees, E. Gerlach

Dept. of Physiology, Univ. of Munich, Federal Republic of Germany

* Dept. of Pharmacology, Univ. of South Carolina, Charleston, U.S.A.

Adenosine (AR) elicits coronary dilatation via an endothelium-mediated process. The vascular endothelium also possesses A_2-receptors which stimulate adenylate cyclase. This activation can be quantified by prelabelling the adenine nucleotide pool with ^{3}H-AR and determining the rate of release of ^{3}H-cAMP. Perfusion of isolated hearts with AR or NECA increases the release of ^{3}H-cAMP from the endothelium into the venous effluent. To establish whether endothelial A_2-receptor activation of adenylate cyclase is the mechanisms leading to vasodilatation by AR, flow responses and endothelial cAMP release were determined in isolated guinea pig hearts in the presence or absence of the P-site agonist 2',5'-dideoxyadenosine (ddA). The capacity of ddA in inhibit adenylate cyclase-mediated effects was independently tested with isoprenaline.

Coronary dilation induced by both AR and NECA was unaffected by the presence of 50 μM ddA, whereas the release of labelled cAMP in response to NECA was reduced by ddA. 50μM ddA also markedly inhibited the positive chronotropic and inotropic actions of isoprenaline. The dissociation elicited by ddA between coronary flow response to AR or NECA and the activation of endothelial adenylate cyclase indicates that the endothelial A_2-receptors are not causally involved in coronary dilation by AR.

AGONISTS AT PURINOCEPTORS: ELECTRICAL AND MECHANICAL EFFECTS IN GUINEA-PIG ISOLATED TRACHEALIS.

J.P.Boyle, J.M.Davies, R.W.Foster, G.P.Morgan, and R.C.Small

Smooth Muscle Research Group, Department of Physiological Sciences, University of Manchester, Manchester M13 9PT, UK.

Mechanical and electrical effects of adenosine and ATP were examined in order to analyse the mechanism of the inhibitory effects of these purines and to assess their ability to mimic the non-adrenergic non-cholinergic (NANC) inhibitory neurotransmitter. Isometric tension recording showed that adenosine (0.01 - 5 mM) and ATP (0.0001 - 5 mM) each caused concentration-dependent suppression of spontaneous tone. Tetra-ethylammonium (8 mM), 4-aminopyridine (5 mM) and procaine (5 mM) each suppressed the shallow foot of the log. concentration-effect curve of ATP but did not depress the upper, steeper part of the ATP curve or any part of that of adenosine. In electrophysiological experiments ATP (< 10 mM) caused relaxation but no membrane potential change. ATP (10 mM) caused a small (7 ± 0.7 mV), transient hyperpolarisation and abolished slow wave activity. Adenosine caused relaxation and increased slow wave amplitude but decreased their frequency without change in resting membrane potential. It is concluded that K^+-channel opening is not a prominent feature of the actions of the purines in the trachealis. The NANC transmitter does not alter resting membrane potential but may suppress slow waves (Boyle et al., 1987). Accordingly, of the two purines, ATP was the better (albeit imperfect) mimic of the NANC transmitter.

Boyle et al., 1987, British Journal of Pharmacology, 90, 733.

EFFECT OF ADENOSINE AND XANTHINE ANALOGS ON RENAL CYCLIC AMP

Richard Coulson, Elizabeth A. Wolszczak & Steven J. Scheinman

VA Medical Center & Departments of Pharmacology and Medicine, SUNY Health Science Center, Syracuse, NY 13210, U.S.A.

In an attempt to demonstrate adenosine receptors in kidney tubule isolated perfused rat kidneys (IPRK) or cultured oppossum kidney (OK) cells were treated with bovine 1-34 parathyroid hormone (PTH) and either adenosine analogs selective for A1 (R-PIA) or A2 (CLA) receptors or xanthine analogs with high (DPX = 1,3-diethyl-8-phenylX) or low (enprofylline) affinity for adenosine receptors. IPRK experiments were conducted without phosphodiesterase inhibitors, OK cell experiments were ± 0.1mM Ro-20-1724. In the IPRK, PIA produced an inhibition of PTH-stimulated cyclic AMP (cAMP) production: 500 μM PIA > 50 μM > 5 μM. [PIA is 80% bound by perfusate albumin, thus 500 μM = 100 μM free PIA, etc]. Conversely, CLA (100 μM) augmented cAMP production by the IPRK treated with PTH. Unlike PIA, CLA also augmented the fractional excretion of Pi. In OK cells, CLA (0.1 - 100 μM) augmented whereas PIA (0.1 - 100 μM) inhibited PTH-stimulated cAMP production, but only in the absence of Ro-20-1724. DPX and enprofylline (50 μM = 5 μM free) augmented PTH-stimulated cAMP production by the IPRK (DPX > enprofylline). DPX, unlike enprofylline, produced a diuresis. In OK cells in the absence of Ro-20-1724, DPX was a much more potent enhancer of PTH-stimulated cAMP than enprofylline. Concentrations producing half maximal stimulation of cAMP were ~5 μM DPX and ~100 μM enprofylline. These data suggest tubule receptors coupled to both PDE and adenylate cyclase activities.

EFFECTS OF ISOPOLAR ISOSTERIC PHOSPHONATE ANALOGUES OF ADENOSINE 5'-DIPHOSPHATE (ADP) ON HUMAN PLATELETS.

N.J. Cusack and C.J. Pettey

Department of Pharmacology, King's College London, Strand, London, WC2R 2LS, U.K.

Adenosine 5'-diphosphate (ADP) induces human platelet aggregation and inhibits stimulated adenylate cyclase, and controversy still exists as to whether these effects require dephosphorylation of ADP. Horak and Barton (1974) found that a nonhydrolysable ADP analogue, adenosine 5'-(α,β-methylene)diphosphonate ($AMPCH_2P$) failed to induce platelet aggregation, and this was attributed to this analogue being not as ionized as ADP at physiological pH.

We synthesised and tested adenosine 5'-(α,β-imido)diphosphonate (AMPNHP), adenosine 5'-(α,β-difluoromethylene)diphosphonate ($AMPCF_2P$) and adenosine 5'-(α,β-dichloromethylene)diphosphonate ($AMPCCl_2P$) to see if restoration of the electronegativity of the diphosphonate chain would provide analogues that behave as ADP itself. AMPNHP was a partial agonist (intrinsic activity 0.5), but $AMPCCl_2P$ was a competitive inhibitor (Schild slope 0.95; K_i 27μM) and $AMPCF_2P$ (Schild slope 0.66) and $AMPCH_2P$ (Schild slope 0.35) were noncompetitive inhibitors of ADP-induced aggregation. None of these analogues inhibited stimulated adenylate cyclase, but $AMPCCl_2P$ and $AMPCF_2P$ antagonised this affect of ADP.

These results suggest that the electronegativity and ionization of the diphosphate chain is of less importance than its conformation, the presence of lone pairs of electrons, or its susceptibility to hydrolysis.

Horak, H. and Barton, P.G., 1974, Biochimica et Biophysica Acta, 471, 373.

INHIBITORY EFFECTS OF ADENOSINE ON MYOCARDIAL CONTRACTILITY: COMPARISON WITH MUSCARINIC RECEPTOR STIMULATION

M. Endoh and H. Kushida

Department of Pharmacology, Yamagata University School of Medicine, Yamagata 990-23, Japan.

Isolated canine left papillary muscles and ventricular trabeculae were electrically driven at a rate of 0.5 Hz at 37°C. The concentration-response curve (CRC) for adenosine (ADE), N^6-phenylisopropyladenosine (PIA) and carbachol (CARB) was determined when the developed tension reached a steady level after administration of various concentrations of isoproterenol (IPN). The extent of inhibition of IPN-induced PIE elicited by ADE or PIA was much influenced by the IPN concentration employed. ADE or PIA inhibited completely the positive inotropic effect (PIE) of 10 nM IPN, decreased the IPE of 30 and 100 nM IPN by 20% of the maximum response to IPN (IPN_{max}), and did not affect significantly the PIE of 1 μM IPN. Adenosine deaminase in a concentration of 1.5 units/ml, which abolished the inhibitory action of ADE did not modify the inhibitory effect of PIA. CARB antagonized the IPN-induced PIE more effectively than ADE or PIA: the maximum inhibition amounted to 50% of IPN_{max}; CARB inhibited the PIE of 1 μM IPN also by 50% of IPN_{max}. The present findings indicate that the lability of adenosine-induced inhibitory action on canine ventricular myocardium is not due to a contribution of endogenously released adenosine, but to less efficacy of agonists acting *via* adenosine receptors. CARB with higher efficacy was capable of inhibiting the PIE elicited by the highest concentration of IPN.

EVIDENCE FOR A1 AND A2 RECEPTORS MEDIATING *IN VIVO* POTENTIATION BY ADENOSINE OF THE SYMPATHOMIMETIC EFFECTS OF NICOTINE.

Gary Evoniuk, Reid W. von Borstel and Kenneth A. Jacobson

National Institutes of Health, Bethesda, MD 20892 USA

Intravenous infusion of subhypotensive doses of adenosine strongly potentiates the sympathomimetic effects of nicotine (40 ug/kg iv) when administered to anesthesized rats. This effect persists after adrenalectomy, blockade of muscarinic receptors, pithing, or decentralization of sympathetic ganglia, but is sensitive to blockade of adenosine receptors with caffeine or 8-p-sulfophenyltheophylline. These data suggest that adenosine receptors within or distal to sympathetic ganglia may be involved in the *in vivo* potentiation of the sympathomimetic effects of nicotine by adenosine. Recently XAC, a 1,3-dipropyl-8-phenylxanthine amine congener has been developed which displays considerable A1 selectivity *in vivo*. Thus, XAC reduces adenosine receptor-mediated bradycardia at doses which have minimal effects on adenosine-mediated hypotension. XAC selectively blocked potentiation by adenosine of the positive chronotropic effect of nicotine at doses which had only minimal effects on enhancement of the hypertensive effect of the drug. It appears that either ganglionic adenosine receptors are the site of the nicotine interaction, and that their subtypes mirror those found at the end organ (i.e. A1 at cardiac and A2 at vascular smooth muscle) or alternatively, that pre- or postsynaptic adenosine receptors within the end organs them-selves are the site of the potentiating effect of the nucleoside.

PURINES ARE REGULATORS OF SKELETAL MUSCLE VASCULAR TONE

L.E. Gustafsson. M. Persson. A. Ohlén and P. Hedqvist.
Department of Physiology, Karolinska Institutet, and National Institute of Environmental Medicine, S-104 01 Stockholm, Sweden.

The effects of adenosine (AR) and the AR receptor antagonists, 8-cyclopentyltheophylline (CPT), 1,3-dipropypl-8-p-sulfophenyl-xanthine (DPSPX), 8-p-sulfophenyltheophylline (8-PSOT) and 8-m-sulfophenyltheophylline (8-MSOT), on vessel diameter and blood flow distribution in rabbit teniussinus muscle were studied using intravital microscopy.

AR dilated both transverse and terminal arterioles dose-dependently. Terminal arterioles were more sensitive than transverse arterioles, leading to a redistribution of blood flow into the muscle tissue. No effect was seen on venular diameters. The vasodilatory effect of AR was competitively antagonized by the AR antagonists, with a potency order (DPSPX>CPT>8-PSOT>8-MSOT) in agreement with their previously known order as AR A_2 antagonists. Furthermore, the AR antagonists per se constricted both transverse and terminal arterioles. Their constrictor potency agreed with their capacity as AR antagonists in this tissue. The antagonist-induced vasoconstriction was not counteracted by phentolamine. The antagonist potency was only in partial agreement with their capacity as AR A_1 and in disagreement with their capacity as AR A_{2b} antagonists. The receptor involved might thus be of the A_{2a} type.

These results suggest that purines have a role in the regulation of resting tone of skeletal muscle arterioles and that AR exerts its effect primarily by directing flow into the muscle tissue proper.

MODULATION BY PURINES AND Ca^{2+}-ANTAGONISTS OF CHOLINERGIC NEUROTRANSMISSION IN GUINEA-PIG ILEUM.

T. Katsuragi, K. Shirakabe, K. Miyamoto and T. Furukawa
Department of Pharmacology, School of Medicine
Fukuoka University, Fukuoka 814-01, Japan

Cholinergic neurotransmission can be modulated by stimulation of presynaptic P_1-purinoceptors. In the present study, to clarify the mechanisms of this purinergic modulation, the effects of purines and a Ca^{2+}-antagonist on the contractile response of the longitudinal preparation from guinea-pig ileum to electrical stimulation were examined. The electrically-evoked twitch contraction was suppressed by adenosine, ATP or α, β-methylene ATP in a concentration dependent manner. The suppression evoked by purine compounds was markedly antagonized by 10^{-4}M theophylline, but were enhanced by $5x10^{-7}$M dipyridamole and by lowering Ca^{2+} in the medium. Nifedipine inhibited the ileal contraction following electrical stimulation, and the inhibition evoked by 10^{-8}M nifedipine was antagonized by theophylline. Bay K 8644 ($3x10^{-8}$M) antagonised the suppression of twitch contractions by α, β-methylene ATP (10^{-6}M) or nifedipine (10^{-8}M). However, the contraction evoked by ACh was virtually unaffected by α, β-methylene ATP (10^{-5}M) and nifedipine (10^{-8}M) indicating that these drugs preferentially inhibited neurotransmission in these concentrations. These findings suggest that the alteration of cholinergic neurotransmission results from Ca^{2+} -antagonistic properties of purine compounds as well as nifedipine.

ADENOSINE SURGE AT BIRTH

H.Lagercrantz, L. Irestedt, M. Runold and A. Sollevi
Nobel Institute for Neurophysiology and Departments of Pediatrics and Anesthesiology (KS), Karolinska Institute, Stockholm, Sweden.

The aim was to determine the plasma adenosine concentrations at birth in relation to the stress of being born (Lagercrantz & Slotkin 1986). Umbilical arterial and venous blood was collected 10-15 s after birth into an ice-cooled solution to prevent degradation of adenosine (Sollevi 1986). Purines and catecholamines (CA) were determined with HPLC.

The mean (+SD) plasma adenosine concentration was 0.61+0.46 μM (n=30) in the umbilical artery and 0.57+0.37 μM in the vein after uncomplicated vaginal deliveries. This is about 2-3 fold higher than in resting adults and also higher than in infants born after elective caesarean section. A linear correlation between arterial hypoxanthine concentration and pH (r=0.81) was found as well as between hypoxanthine and CA.

We also found that adenosine markedly modulated circulatory and metabolic effects of CA in neonatal piglets and we speculate that adenosine might protect the fetus during birth stress.

Lagercrantz, H. & Slotkin, T, 1986 Scientific American, 254, 100.
Sollevi, A., 1986 Progress in Neurobiology, 27, 319.

ADENOSINE AND NEONATAL APNEOA

H. Lagercrantz, M. Runold and B. Fredholm
Nobel Institute for Neurophysiology, Departments of Pharmacology and Pediatrics (KS), Karolinska Institute, Stockholm, Sweden.

Hypoxia causes a sustained decrease of breathing in the newborn mammal after transient stimulation. We wished to investigate to what extent adenosine is involved in the mediation of this so-called hypoxic depression of breathing in the newborn. Newborn rabbit pups were anaesthetized with urethane and their breathing monitored with pheumotachography. The effect of 6% O_2 in N_2 was tested.

In the control series 10 out of 16 pups became completely apnoeic in 9 min. In another series of pups pretreated with theophylline (25 mg/kg b.w., i.p.) only one out of 8 pups became apnoeic in 15 min. In a third series of animals we attempted to increase the endogenous concentration of adenosine by pretreatment with dipyridamole. This enhanced the hypoxic depressenteffect.

These findings suggest that adenosine is involved in the mediation of hypoxia-induced apneoa in newborns. Besides respiratory depression, adenosine has been reported to induce hypotonia, hypothermia and hypotension which are also typical symptoms seen after birth asphyxia.

Runold, M., Lagercrantz, H., Fredholm, B., 1986, J.Appl.Physiol. 61, 255.

ROLE OF ADENOSINE IN REGULATION OF MESENTERIC ARTERIAL BLOOD FLOW

W.W. Lautt, D.J. Legare, L.A. D'Aleo, J.E. McQuaker, L.K. Lockhart
Hepatorenal Research Unit, Department of Pharmacology and Therapeutics, Faculty of Medicine, University of Manitoba, Winnipeg, Manitoba, Canada, R3E 0W3.

The superior mesenteric artery in fasted cats shows pressure-flow autoregulation that is blocked by 8-phenyltheophylline (8-PT) in a dose-dependent manner. 8-PT did not alter the vasodilator response to isoprenaline while completely blocking the response to exogenous adenosine. The data are compatible with the hypothesis that classical autoregulation of gut blood flow is mediated by adenosine and is not myogenic. In support, elevations in portal venous pressure over physiological ranges did not cause a myogenic constriction. Post-occlusive hyperemia was also antagonized by 8-PT.

Exogenous adenosine in doses that do not cause maximal vasodilation produced dose-related inhibition of vasoconstrictor responses to noradrenaline(NA) and sympathetic nerve stimulation with no evidence for presynaptic neuromodulation. Vasoconstriction induced by 3 and 9Hz nerve stimulation and by low and high doses of NA were antagonized to similar degrees by adenosine at each of three doses of adenosine. Vascular escape from neurogenic constriction and post-stimulatory hyperemia were not altered by adenosine of 8-PT administration.

Thus, in the feline small intestine, adenosine is responsible for the vasodilation induced by decreased blood pressure (autoregulation) and post-occlusive hyperemia; vasoconstriction induced by sympathetic nerves and NA is antagonized by exogenous adenosine in concentrations that produce moderate vasodilation.

CHARACTERIZATION OF NEURAL AND CARDIAC ADENOSINE RECEPTORS BY AGONIST AND ANTAGONIST RADIOLIGANDS.

E. Leung[1], K.A. Jacobson[2] and R.D. Green[1]
Department of Pharmacology, University of Illinois, Chicago, IL[1] and Laboratories of Chemistry and Bioorganic Chemistry, NIH, Bethesda, MD[2]

Several laboratories have utilized agonist radioligands ([3H]R-PIA, [125I]HPIA and [125I]ABA) to study cardiac adenosine receptors. These studies have shown that the cardiac adenosine receptor is present in low concentrations compared to the CNS and that the cardiac receptor has the characteristics of the CNS A1 receptor.

The ability of [3H]XAC, a new high affinity A1 receptor antagonist, to bind to preparations of chick and bovine myocardium has been tested. [3H]XAC bound to receptors in membranes of chick myocardium and cerebellum with similiar affinities ($K_d \sim$ 5-10 nM) and to receptors in membranes from bovine heart and cerebral cortex with similiar, but much higher, affinities ($K_d \sim$ 0.1 nM). Comparisons of the binding of [3H]XAC, [3H]R-PIA and [125I]ABA to cardiac and brain receptors from both species showed that the agonist ligands and the antagonist ligand bound to similiar numbers of sites in brain membranes but that the number of sites in cardiac membranes that were detected with the agonist ligands were considerably less than the numbers of sites detected by [3H]XAC. The finding that [3H]XAC has similiar affinities for adenosine receptors in the heart and in the CNS supports the earlier conclusion that the cardiac receptor is similiar to the CNS A1 receptor.

THE ROLE OF ADENOSINE IN SYNAPTIC DEPRESSION EVOKED BY A BRIEF PERIOD OF HYPOXIA IN THE ISOLATED RAT SPINAL CORD.

H.G.E. Lloyd, I. Spence and G.A.R. Johnston
Department of Pharmacology, University of Sydney, NSW 2006, Australia.

The observation that cessation of superfusion of hemi-sected spinal cords of newborn rats (7-11 days old) results in a rapid depression of the monosynaptic reflex (MSR) - which can be recorded extracellularly after stimulation of the corresponding dorsal root - led to an investigation of the mediator(s) of this depression. A 4 min period of superfusion stop-flow reduced the MSR rsponse to 32±8% (mean ± sem; n=4) of control: this was readily reversible, 100% recovery occurring within 3 min of reperfusion. This observed depression was completely absent in the presence of an adenosine antagonist 8-cyclopentyltheophylline (8-CPT; 10^{-6}M). In a second series of experiments tissue was superfused with anoxic medium equilibrated with 95% N2, 5% CO2 for 4 min. The MSR was reduced to 32±3% (n=8) which recovered to 100%, although less readily (15 min of superfusion using Krebs equilibrated with 95% O2, 5% CO2 was required). In the presence of 8-CPT (10^{-6}M) 4 min of anoxic perfusion decreased the MSR to 90±4% (n=4) of control. These results, together with the observations that: (1) exogenous adenosine, when applied to the spinal cord, produces a rapid but reversible depression of the MSR and (2) hypoxia is a potent stimulus for adenosine release, provide strong evidence for a critical role of adenosine in mediating synaptic depression evoked by a brief period of hypoxia.

COMPARISON OF THE CARDIAC EFFECTS OF ADENOSINE (ADO) ANALOGS IN RAT HEART AND THEIR RECEPTOR BINDING ACTIVITY IN RAT BRAIN.

H.H.Oei*, G.R.Ghai*, H.C.Zoganas*, G.A.Stone*, L.Sanathanan*, M.Williams and M.B.Zimmerman*.
Research Department, Pharmaceuticals Division, CIBA-GEIGY Corp. Summit, NJ 07901.

ADO and its analogs produce physiological effects on cardiac function by interaction with ADO receptors. Since direct measurement of cardiac ADO receptors has been limited by their low density, receptor activity at the biochemical level has been determined in brain tissue. Thus, the effects of the ADO analogs R-N^6-phenylisopropylADO (R-PIA), 5'-N-ethyl-carboxiamidoADO (NECA) and 2-chloroADO (2-CADO) in reducing heart rate and increasing coronary flow in isolated, perfused rat heart were compared with their activities at rat brain A_1 and A_2 receptors using ligand binding. At the brain A_1 receptor (ligand=[^{3}H]CHA), IC_{50} values for R-PIA, NECA and 2-CADO were 1.6, 6.3 and 14 nM, a relative activity of 1:4:9. In decreasing heart rate, an A_1-mediated response, EC_{25} values of 9.3, 57 and 120 nM were obtained, a potency relationship of 1:10:13. At rat striatal A_2 receptors (ligand=[^{3}H]NECA), the activities of R-PIA, NECA and 2-CADO were 241, 12 and 86 nM (20:1:7), while EC_{25} values for increasing coronary blood flow, and A_2 response, were 61, 5 and 50 nM (13:1:10). The similarity in the relative relationships suggests that activity in brain tissue is predictive of physiological activity in cardiac tissue and that similarities exist in A_1 and A_2 receptors in the two tissues.

COMPARATIVE EFFECTS OF 5´-N-ETHYLCARBOXAMIDO-ADENOSINE (NECA) AND OF L-PHENYLISOPROPYLADENOSINE (L-PIA) ON ISOLATED GUINEA-PIG HEART ATRIA

M. Prostran and V.M. Varagić

Department of Pharmacology, Faculty of Medicine, P.O. Box 662, University of Belgrade, Belgrade, Yugoslavia

In the present experiments, it was of interest to compare the effects of L-PIA and NECA on the isolated spontaneously beating guinea-pig heart atria. Both L-PIA (35-350 nM) and NECA (2.16-22 nM) produced a concentration-dependent depression of both the isometric contractility and the atrial rate. The depressant effects of both substances were potentiated in the presence of verapamil (73nM). The addition of calcium (1.8mM), aminophylline (AMPh, 32 μM) and isoprenaline (24 nM) antagonized the negative inotropic effects of both L-PIA and of NECA. In the presence of AMPh (6.4 μM), the dose response curves for both L-PIA and NECA were shifted to the right. Our results indicate a possible antagonism between L-PIA and NECA on the one hand, and of Ca on the other hand, on the isolated heart atria of the guinea-pig, and suggest a competitive antagonism between L-PIA or NECA and AMPh at adenosine receptor sites.

PURINERGIC RECEPTORS IN THE BRAINSTEM MEDIATE HYPOTENSION AND BRADYCARDIA

D. Robertson, C.-J., Tseng, M. Appalsamy, I. Biaggioni
Departments of Medicine and Pharmacology, Vanderbilt University School of Medicine, Nashville, Tennessee 37232

Adenosine acts at many sites to modulate neuronal activity. The purpose of this study was to investigate a possible role for adenosine as a neuromodulator of brainstem cardiovascular control. Microinjections of adenosine (0-2.3 nmol) were made stereotaxically into various brainstem sites. Injection of adenosine into the nucleus tractus solitarii (NTS) produced dose-related decreases in heart rate, systolic and diastolic blood pressures. Maximal changes occurred at 90 seconds after injection. Injection into the area postrema also produced decreased heart rate, systolic and diastolic blood pressures. No significant effect occurred following injection into the C1 area. Adenosine 5'-triphosphate (ATP) and its analogue, β,γ-methylene ATP also produced dose-related and potent vasodepressor and bradycardia effects in the NTS. Injection of 1,3-dipropyl-8-p-sulfophenylxanthine (0.92 nmol), a potent adenosine receptor antagonist, produced no effect itself, but completely abolished for 45 minutes the actions of further injections of adenosine and ATP (but not L-glutamate) in both the NTS and area postrema. Thus, NTS and area postrema injections of adenosine decrease blood pressure and heart rate in anesthetized normotensive rats through adenosine receptors located in these areas. These findings support a role for endogenous adenosine as a central modulator in cardiovascular control.

DISTRIBUTION OF ADENOSINE RECEPTORS IN THE RENAL CORTEX.

W. Schütz, C. Nanoff and M. Freissmuth
Institute of Pharmacology, University of Vienna, Wahringer Str. 13a, A-1090 Vienna, Austria

Adenosine (AR) may regulate renin secretion, renal blood flow and epithelial electrolyte secretion, suggesting the presence of AR receptors on renal glomuruli, microvessels and tubules. In all three fractions of rabbit renal cortex (isolated by collagenase dispersion and purification on density gradient) A_2 receptors were demonstrated on the bases of adenylate cyclase (AC) stimulation by AR analogs (NECA, R-PIA) while methylxanthines acted as antagonists. Binding studies with [^{125}I]-HPIA revealed a low density of A_1 receptors in the glomerular and microvessel fractions. These were altered into a low affinity state by guanine nucleotides, indicating coupling to a G-protein. The potency rank order of competitors for [^{125}I]-HPIA binding was R-PIA>NECA>S-PIA. However, no inhibition of AC activity by these analogs was detectable. This failure may be due to the difficulty of demonstrating AC inhibition (A_1-site) in the presence of activation (A_2-site), or to A_1-receptor occupation resulting in signal transduction not coupled to AC.

The tubular A_2-receptor may be the target for cAMP-dependent stimulation of transepithelial ion transport by AR, while the glomerular and microvessel A_1 receptor may be the target site of AR-induced inhibition of renin secretion, recently shown (Rossi et al. 1987) to be a G-protein-dependent process. the exact function of A_2-receptor observed on glomuruli and microvessels remain unclear. Rossi, N.F., Churchill, P.C., Churchill, M.C. 1987, Life Sciences, 40, 481-487.

ENDOTHELIUM-DEPENDENT VASOCONTRACTIONS OF CANINE CEREBRAL ARTERY

Hachiro Usui, Kazuyoshi Kurahashi, Hiroaki Shirahase and Motohatsu Fujiwara
Department of Pharmacology, Faculty of Medicine, Kyoto University, Kyoto 606, Japan

The contractions of canine cerebral arteries induced by ACh, noradrenaline and nicotine are not seen in "rubbed" preparations from which the endothelium had been removed. Similarly, the contractions to be induced by arachidonic acid and a calcium ionophore (A-23187) also require an intact endothelium. All such contractions of intact vessels were inhibited by cyclooxygenase inhibitors (aspirin, indomethacin), by a thromboxane A_2 synthetase inhibitor (OKY-046) and by a thromboxane A_2 antagonist (ONO-3708). These results suggest that endothelium-dependent contractions are caused by stimulation of thromboxane A_2 production and its subsequent release from the endothelium. ATP, but not adenosine, induced endothelium-dependent contraction, and this was unaffected by an adenosine antagonist (8-phenyltheophylline), but inhibited by aspirin, OKY-046, and ONO-3708. It is concluded that endothelium-dependent contractions of canine cerebral artery result from the metabolism of arachidonic acid.

INDIRECT NEUROMODULATION BY ADENOSINE 3'-PHOSPHONUCLEOTIDES IN GUINEA-PIG ILEUM.

C.U. Wiklund, N.P. Wiklund & L.E. Gustafsson
Department of Physiology, Karolinska Institutet and National Institute of Environmental Medicine, Stockholm, Sweden.

Acetyl-coenzyme A (AcCoA), adenosine 3',5'-diphosphate (3'5'ADP) and adenosine 3'-monophosphate (3'AMP) inhibited contractile responses to transmural nerve stimulation in guinea pig ileum longitudional muscle via a presynaptic action. The effects of the nucleotides were antagonized by 8-p-sulfophenyltheophylline, a competitive adenosine receptor antagonist. At low antagonist concentrations, dose-response curves for the nucleotides were displaced to the right in a parallel fashion. At higher antagonist concentrations the nucleotides did not reach maximal effects, suggesting non-competitive antagonism. These findings are compatible with an action by the nucleotides via metabolites. The effect of 3'AMP was abolished by 10 min applications of adenosine deaminase at 1.0 U ml^{-1} suggesting an action by 3'AMP via breakdown to adenosine.

In summary the results suggest that adenosine 3'-phospho-nucleotides cannot per se act on adenosine receptors. 3'AMP is likely to act via breakdown to adenosine. AcCoA and 3'5'ADP probably act via breakdown to adenosine or a 5'nucleotide such as AMP or ADP.

ADENOSINE SMOOTH MUSCLE STIMULATION AND ENHANCEMENT OF ADRENERGIC TRANSMISSION IN GUINEA PIG PULMONARY ARTERY

N.P. Wiklund[1], B. Cederqvist[1] & L.E. Gustafsson[1,2]

Department of Physiology[1], Karolinska Institutet and National Institute of Environmental Medicine[2], Stockholm, Sweden

Transmural stimulation of short strips of guinea pig pulmonary artery elicited contractile responses, abolished by tetrodotoxin or phentolamine. Adenosine, NECA or R-PIA usually enhanced contractile responses to transmural nerve stimulation and augmented the basal smooth muscle tone. However, in a few preparations inhibition of the nerve-induced contractile responses was noted. The effects of adenosine and its analogues were antagonized by 8-p-sulfophenyltheophylline.

R-PIA or NECA enhanced contractile responses induced by noradrenaline, whereas S-PIA was virtually inactive, suggesting postjunctional enhancement of neurotransmission via adenosine A_1 receptors.

R-PIA and NECA inhibited the stimulus-evoked release of [^{3}H]-noradrenaline ([^{3}H]-NA). However, total [^{3}H]-NA overflow was increased by NECA whereas R-PIA inhibited the total [^{3}H]-NA overflow. Desipramine and normetanephrine abolished the increase in total [^{3}H]-NA overflow produced by NECA.

In isolated spiral strips of guinea pig pulmonary artery, the effects of adenosine and its analogues were mainly due to actions at post-junctional stimulatory A_1 receptors but actions at prejunctional inhibitory A_1 and stimulatory A_2 receptors were also evident. These findings may suggest a role for adenosine in the regulation of pulmonary blood flow.

6-NITRO-2-PHENYLISATOGEN ENHANCES ADENOSINE BUT NOT 2-CHLOROADENOSINE EFFECTS IN GUINEA-PIG ILEUM

D. Wood

Postgraduate School of Pharmacology, University of Bradford, Bradford, West Yorkshire, U.K.

6-Nitro-2-phenylisatogen (6N2PI) has been reported (Wood,1986) to enhance the inhibitory effect of adenosine (AD) upon cholinergic transmission in guinea-pig ileum. In this study, the nature of the enhancement has been explored by comparing the effect of 6N2PI on responses to AD and to the stable analogue, 2-chloroadenosine (CADO).

Segments of ileum received transmural stimulation at 0.1Hz whilst longitudinal tension changes were recorded. Cumulative concentration-response curves for the inhibition of twitch by AD or CADO were constructed following sequential exposure of the tissue for 0.5h to vehicle and to increasing concentrations of the isatogen. Reciprocal dose ratios (RDR) were calculated as the ratio:control $EC_{50} \div EC_{50}$ in presence of isatogen. 6N2PI, at a concentration (10μM) producing a significant increase in the RDR for AD (10.3±1.0, n = 8) did not significantly affect responses to CADO (RDR 0.8±0.1, n = 6). In similar experiments, dipyridamole (0.1μM) enhanced AD responses (RDR 18.8±3.0,n = 8) whilst being ineffective to enhance CADO responses (RDR 1.4±0.2,n = 4). Thus, 6N2PI may share with dipyridamole the ability to inhibit inactivating mechanisms for adenosine, since both these compounds enhance responses to AD but not to CADO.

Wood, D., 1986, British Journal of Pharmacology, 87, 186P.

RENAL VASCULAR EFFECTS OF ADENOSINE IN ANESTHETIZED DOGS

T. Yukimura, K. Mira, K. Ito and K. Yamamoto
Department of Pharmacology, Osaka City University Medical School
1-Asahimachi, Abeno, Osaka 545, JAPAN

Renal vascular responses to adenosine (AR) are unique (Ueda et al.1974). Intrarenal infusion (i.r.a) of AR (0.05-1.0 μmol/min) in anaesthetized dogs caused an initial decrease followed by sustained increase in renal blood flow without any change in blood pressure. These effects were not modified by pretreatment with phenoxybenzamine, propranolol, atropine (all 1 mg/kg) or the angiotensin converting enzyme inhibitor, SA-446 (0.1 mg/kg) but were significantly eliminated by isobutylmethyl xanthine (IBMX 0.5-2. μmol/min i.r.a.) pretreatment, indicating that the renal vascular action of AR are mediated through an AR receptor but not through the autonomic nervous system or the renin-angiotensin system. Since renal vasoconstriction by AR was eliminated by verapamil (5-30 ug/min i.r.a.), enhancement of calcium channel-linked AR receptor may be involved in the renal vasoconstriction by AR. Dilazep potentiated both the renal vasoconstrictive and vasodilatory actions of AR and itself caused renal vasodilation in the absence of vasoconstriction. This vasodilatory effect of dilazep was inhibited by IBMX, suggesting an action on renal vasculature mediated through AR receptors (Yukimura et al. 1986).

Ueda, J. et al., 1974, Osaka City Medical Journal, 20, 33-350.
Yukimura, T. et al., 1986, Journal of Cardiovascular Pharmacology, 8, 649-655.

INDEX

H

I

L

M

N

V

X